Human Well-Being Research and Policy Making

This series includes policy-focused books on the role of the public and private sectors in advancing quality of life and well-being. It creates a dialogue between well-being scholars and public policy makers. Well-being theory, research and practice are essentially interdisciplinary in nature and embrace contributions from all disciplines within the social sciences. With the exception of leading economists, the policy relevant contributions of social scientists are widely scattered and lack the coherence and integration needed to more effectively inform the actions of policy makers. Contributions in the series focus on one more of the following four aspects of well-being and public policy:

- Discussions of the public policy and well-being focused on particular nations and worldwide regions
- Discussions of the public policy and well-being in specialized sectors of policy making such as health, education, work, social welfare, housing, transportation, use of leisure time
- Discussions of public policy and well-being associated with particular population groups such as women, children and youth, the aged, persons with disabilities and vulnerable populations
- Special topics in well-being and public policy such as technology and well-being, terrorism and well-being, infrastructure and well-being.

This series was initiated, in part, through funds provided by the Halloran Philanthropies of West Conshohocken, Pennsylvania, USA. The commitment of the Halloran Philanthropies is to "inspire, innovate and accelerate sustainable social interventions that promote human well-being." The series editors and Springer acknowledge Harry Halloran, Tony Carr and Audrey Selian for their contributions in helping to make the series a reality.

Enrico Ivaldi • Andrea Ciacci

Smart Sustainable Cities and Knowledge-Based Economy

Policy Implications

Springer

Enrico Ivaldi
Department of Political
and International Sciences
University of Genoa
Genoa, Italy

Andrea Ciacci
Department of Economics and Business Studies
University of Genoa
Genoa, Italy

ISSN 2522-5367 ISSN 2522-5375 (electronic)
Human Well-Being Research and Policy Making
ISBN 978-3-031-25037-8 ISBN 978-3-031-25038-5 (eBook)
https://doi.org/10.1007/978-3-031-25038-5

This Springer imprint is published by the registered company Springer Nature Switzerland AG
The registered company address is: Gewerbestrasse 11, 6330 Cham, Switzerland

Foreword

With the growing world population and rural–urban wage gap coupled with people's preference for urban livelihoods, the spaces occupied by the urban sector have tremendously increased at the global level. Using a simple measure of urbanization as the ratio of urban people to the total population, the World Bank data show that the share of urban population of the world was 33.6% in 1960 and it bounced to 56.8% in 2021. In terms of figures, the total number of urban people was 1.02 billion in 1960 and it jumped up by four times to 4.43 billion in 2021. The data on the shares of urban population across some of the developed and developing countries show that the USA has 83 % of its total population living in the urban area in 2021 while the UK, Germany, and Italy have 84%, 78%, and 71%, respectively. The European Union has 75% of its population in the urban area. According to the statistics from the world's so-called developing as well as highly populous countries, China, India, and Brazil, the corresponding shares are 63%, 35%, and 84% in 2021. Brazil is now the world's highly urbanized country.

The huge figures for the urban population are encouraging, but it is not viable so far as the sustainable development goals are concerned. Many of the world's highly populated cities are contributing hugely to the accumulation of atmospheric greenhouse gases. The rate and number of airborne and water-borne diseases are increasing day by day. As policies against these evil syndromes, the provisioning of so many urban amenities, utility services, etc. are needed for making the people smart. Further, to be a smart person in the urban area, one will have to have smart knowledge. The urban authorities in smart cities usually provide knowledge-based capital for the smooth running of the livelihoods on the one hand and maintaining sustainability of the urban space on the other hand. The smart city adopts strategies to establish relationships between material infrastructures and those who inhabit them. They manage resources intelligently, promote sustainable development, and aim for energy self-sufficiency. A smart city is always attentive to the quality of life and the needs of citizens; it keeps pace with technological innovations and the digital revolution.

One such zone where these smart practices are largely operated is Europe. The book titled *Smart Sustainable Cities and Knowledge-Based Economy: Policy Implications* by Dr. Enrico Ivaldi and Dr. Andrea Ciacci has attempted to examine both the regional system of entrepreneurship and the quality of life and well-being at the smart city levels in the continent. Besides, it provides synthetic indexes to assess the relationship between the perceived quality of life and entrepreneurship in Europe, in particular. These interrelated relationships between knowledge-based economy and smart sustainable cities create patterns of mutual influence like a circuit of knowledge transmission. In this regard, as the authors mentioned, entrepreneurship plays a crucial role since it allows for the commercialization of knowledge. Further, the book discusses how digital platform-based entrepreneurial ecosystems stimulate the knowledge flows within a smart sustainable city and play a relevant role in its growth. The authors weigh upon digital platforms which raise the cities' smartness and sustainability by facilitating knowledge exchanges with partners and improving knowledge management processes inside the firm. Finally, the book prescribes public–private partnerships which would incentivize digital platform adoption for collaborative purposes.

The contents of the book are of great relevance in today's world so far as cost minimization in the urban livelihoods and sustainable development in the urban areas are concerned. I expect that the readers in the related fields will be immensely benefitted by following the book for their further contributions to the existing literature. I congratulate Dr. Ivaldi and Dr. Ciacci, and the publisher, for presenting such a good title before the readers.

Department of Economics, Vidyasagar University, Midnapore, West Bengal, India

Ramesh Chandra Das

Introduction[1]

By 2030, cities will be larger and have more inhabitants—this will have an unprecedented impact on their infrastructure endowment and resources. The world's ten most populous cities (which will attract 35% of the population increase) are estimated to be Tokyo (37.2 million), Delhi (36.1 million), Shanghai (30.8 million), Mumbai (27.8 million), Beijing (27.7 million), Dhaka (27.4 million), Karachi (24.8 million), Cairo (24.5 million), Lagos (24.2 million), and Mexico City (23.9 million) (UN, World Urbanization Prospects: The 2018 Revision, 2018, 2022). An urbanization phenomenon of this intensity concentrates most of the global wealth (about 80% of GDP) in metropolises but also inevitably problems. These include worsening air pollution, chaotic urban sprawl (such as the development of slums in some areas of the world), inadequate and overburdened infrastructure and services, and excessive consumption of energy resources (more than half of the global resources are consumed by urban centers) (World Bank, 2020). Around the world, many cities have realized that their performance and reductions in criticality depend not only on the size of their population, production facilities, and physical infrastructure but, even more, on the availability and communication of knowledge and their social and intellectual capital. Smart cities are distinguished from their more technological counterparts by the effective application of knowledge and technology in a coherent framework, benefiting from the growing importance of information and communication technologies of social and environmental capital, and are also becoming an important marketing concept and tool. A city with high intellectual capital has an advantage in the transition: local government must promote, as mentioned above, good policies aimed at developing sustainable and ecological practices. Similarly, private companies, if embedded in a knowledge-based environment, have the opportunity to take advantage of the favorable conditions that are created: the use of highly

[1] The book is the result of the joint effort of the two authors. However, Chapter 1 is to be attributed to Andrea Ciacci, Enrico Ivaldi, and Marianna Bartiromo. Chapter 3 is to be attributed to Andrea Ciacci. Chapter 4 is to be attributed to Enrico Ivaldi. Chapters 2, 5, and 6 are the joint effort of the two authors, as well as the Introduction and Conclusions.

skilled and educated workers in rethinking the production and distribution process results in the achievement of important sustainability goals. Although there are different definitions of smart cities, the common intents of a smart city, which cut across all definitions, include improving the quality of life of citizens, enhancing the competitiveness of cities, integrating the functioning of urban infrastructure and services, and promoting the well-being, inclusion, participation, environmental quality, and smart development of a city (Batra S., 2014). These intents seem to be in line with those of knowledge cities, which "are designed specifically to encourage the cultivation of knowledge and its sharing among citizens" and which cultivate and harness the knowledge resources residing in the citizens for value creation (Penco et al., 2020).

To become smart, a city must place knowledge at the center of its strategic vision: such an approach to urban planning falls under the term "knowledge-based urban development" (KBUD) (Yigitcanlar, 2010). The KBUD concept allows cities to make the transition to more sustainable and inclusive places: a city goal of attracting and retaining human and intellectual capital has a positive effect on the development of their territory at an economic, social, and cultural level (Pancholi et al., 2015).

The key elements for the transformation of a knowledge-based city into a smart and sustainable city are three, i.e., policies, the business fabric, and human capital. A city with high intellectual capital has an advantage in the transition: local government must promote, as mentioned above, good policies aimed at developing sustainable and environmentally friendly practices. Similarly, private companies, if embedded in a knowledge-based environment, have the opportunity to take advantage of the favorable conditions that are created: the use of highly skilled and educated workers in rethinking the production and distribution process results in the achievement of relevant sustainability goals (Ivaldi et al., 2020).

This study aims to contribute to the debate on sustainable smart cities and the knowledge-based economy from a quality of life and urban entrepreneurship perspective while providing insights, especially for urban policymakers.

In Chapter 1, a review is proposed in order to retrace the history and evolution of cities according to knowledge and sustainable paradigms, as well as the determinants that have progressively accelerated their development. In addition, the sectors most involved in this change are identified and an in-depth analysis is provided. Finally, an excursus of the most crucial challenges that still await cities so that they can decisively define an intellectual and green identity is provided.

In Chap. 2, authors focus on the knowledge-based economy and smart sustainable city (SSC) to define the nature of their interrelationships. The authors start exploring the foundations of the knowledge spillover theories. Therefore, the authors identify the complementarities between the knowledge economy and the SSC. It emerges that the SSC is a functional urban pole to accumulate more knowledge and attract creative people and knowledge workers. This tendency culminates with a mutual self-reinforcing process where knowledge spillovers and clusters of firms synergistically contribute to creating value for the city. In this chapter, the authors define the assumptions to achieve a higher level of quality of life in the city and, in particular, for workers. Knowledge spillovers shape the city's structural, industrial, and social environment and influence urban development.

Chapter 3 aims to discuss how entrepreneurial ecosystems based on digital platforms favor the flows of knowledge toward the overall growth of the smart sustainable city. Digital platforms are information technologies where the great potential of innovativeness lies. By ensuring a constant connection among firms that put in common their knowledge and innovative processes, digital platforms are the foundations to create innovation able to raise the smartness and sustainability levels of cities. This chapter draws on the existing literature about digital platforms, knowledge and strategic management, smart sustainable cities, and entrepreneurship. It elucidates the most relevant literature for this purpose. Therefore, it aligns with different theoretical frameworks.

Chapter 4 provides an example of using index construction techniques to measure the smart sustainable city (SSC) as a multidimensional object. SSC is a complex entity composed of multiple dimensions. SSC works correctly when its composite systems work in a symbiotic way. For this reason, it is relevant to constantly monitor the performance of the different dimensions of an SSC. Monitoring is a working method to identify strengths, weaknesses, risks, and problems, prevent cascade effects, formulate development strategies, and, more in general, govern a city. This chapter develops monitoring of SSC multiple dimensions by employing four index construction methods (e.g., Peña distance, Mazziotta and Pareto, the sum of standardized indicators, and average height). The results reveal that rankings from the different indexes are similar. Constructing monitoring through these methods is not distortive since various indexes confirm the same results. From a methodological viewpoint, the choice of the proper index should depend on the conceptual-theoretical consistency with the framework to be analyzed.

Chapter 5 contains a few suggestions for policy to enhance human well-being in the city. The SSC should grant authentic human-scale relations, which are often in jeopardy in the urban way of life. The chapter identifies which policies are most successful on the basis of the different critical issues and priorities that characterize the various urban fabrics.

Chapter 6 focuses on the theoretical strand that defines the cities of the future according to the meanings of "knowledge" and "sustainability." In this regard, we devote special attention to the evolution of cities. Each new city paradigm ("digital," "smart," "sustainable," "resilient," "15 minutes," and "circular") is treated individually in relation to its evolutionary stage.

At the end of each chapter, we present a box showing an example of an SSC associated with the main insights of the chapter. The provided examples are differentiated on a geographical basis.

Finally, in the Conclusions, we summarize the hallmarks of the SSCs and put forward a few suggestions for policy in order to enhance the quality of life and develop entrepreneurship in the cities. Opportunities and threats that prevent the completion of the transition to a knowledge-based economy within an SSC are, finally, examined in depth.

The book primarily addresses public policymakers on the importance of smart and sustainable cities and their relationship to the knowledge-based urban economy. Overall, this book aims to answer the following questions: how can we implement policies and programs that could make cities "smart" and boost their knowledge-based development? How can we make our constituents happy, reducing inequality and improving the environment where they live and work?

We would like to try to suggest a new approach to creating SSCs, not only in metropolises. We are convinced that European countries must also take on this new and exciting task as a possible path to recovery after the current crisis. Sincerely, we hope these pages can be a valuable contribution.

References

Batra S. (2014) Smart or Knowledge Cities: Which are more relevant for India? 21st Thinkers Writers Forum, Minimum Government Maximum Governance 51-67 Skoch Development Foundation

Ivaldi, E., Penco, L., Isola, G., and Musso, E. (2020). Smart sustainable cities and the urban knowledge-based economy: A NUTS3 level analysis. *Social Indicators Research*, 150(1), 45-72.

Pancholi, S., Yigitcanlar, T., and Guaralda, M. (2015). Public space design of knowledge and innovation spaces: learnings from Kelvin Grove Urban Village, Brisbane. *Journal of Open Innovation: Technology, Market, and Complexity*, 1(1), 13.

Penco, L., Ivaldi, E., Bruzzi, C., and Musso, E. (2020). Knowledge-based urban environments and entrepreneurship: Inside EU cities. *Cities*, 96, 102443.

UN. (2018). *World Urbanization Prospects: The 2018 Revision.* UN Department of Economic and Social Affairs Population Division.

UN. (2022). *Make cities and human settlements inclusive, safe, resilient and sustainable*. Retrieved from https://sdgs.un.org/goals/goal11

Yigitcanlar, T. (2010). Making space and place for the knowledge economy: knowledge-based development of Australian cities. *European Planning Studies*, 18(11), 1769-1786.

WorldBank. (2020). *Urban Development*. Retrieved from The World Bank: https://www.worldbank.org/en/topic/urbandevelopment/overview#1

Contents

Chapter 1
The History and Evolution of Cities in Terms of the Sustainability and Knowledge-Based Economy Sectors

Andrea Ciacci, Enrico Ivaldi, and Marianna Bartiromo

Abstract In this chapter, a review is proposed in order to retrace the history and evolution of cities according to knowledge and sustainable paradigms. This chapter also identifies the determinants that have progressively accelerated smart sustainable city development. Therefore, the sectors most involved in the transition toward a smart sustainable knowledge city model are recognized and analyzed. Finally, the chapter ends with an excursus of the most crucial challenges that still await cities so that they can decisively define an intellectual and green identity.

Keywords Smart sustainable city · Knowledge-based economy · Sustainability · Digitalization

1.1 Birth, Development, and Transformation of Smart Cities toward the Sustainable Model

Nowadays, more and more people live in cities than in the past. This increasing urbanization has made it necessary for cities to become increasingly smart and sustainable in order to respond effectively and efficiently to the growing complexity of the urban environment (Ciacci et al., 2021; Eremia et al., 2017). Urbanization, in fact, "is a complex socio-economic process that transforms the built environment, converting formerly rural into urban settlements, while also shifting the spatial distribution of a population from rural to urban areas" (United Nations et al., 2019, p. 3). According to United Nations forecasts, the world population between 2015 and 2050 will increase by 32% while the urban population will increase by 63%. Current estimates also suggest that by 2030 more than 60% of the population will live in cities with significant increases in Africa, Asia, and Latin America (United Nations et al., 2014). Figure 1.1 shows the annual growth in world population and urban population. As can be seen, an increase in total population corresponds to an increase in urban population.

This increasing urbanization stems from different needs "on one hand to the migration of population from rural areas to cities-in hope for a better life (for jobs, education, medical care, access to culture, etc.), and on the other hand to the

E. Ivaldi, A. Ciacci, *Smart Sustainable Cities and Knowledge-Based Economy*, Human Well-Being Research and Policy Making,
https://doi.org/10.1007/978-3-031-25038-5_1

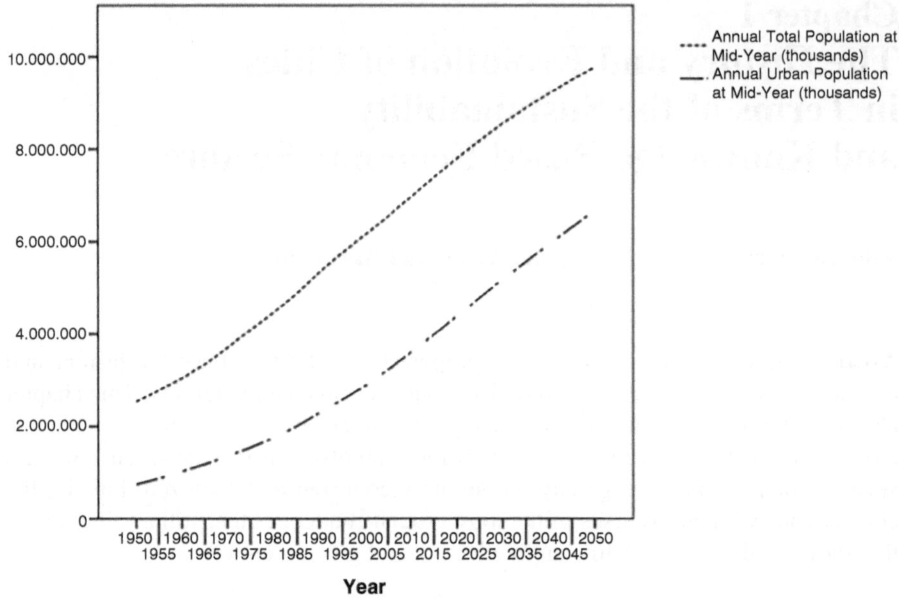

Fig. 1.1 Annual growth in world population and urban population (Source: United Nations)

migration from poor countries or under social and military conflicts toward the industrialized countries" (Eremia et al., 2017, p. 13). The consequence of this will be a significant expansion of the urban environment and the emergence of new cities (Toli & Murtagh, 2020). As cities are born and expand, their consumption will also increase in fact, it is estimated that material consumption related to cities will increase from 40 billion tons in 2010 to about 90 billion tons by 2050 (Hoballah et al., 2012; Toli & Murtagh, 2020). Therefore, it can be understood how this phenomenon, in addition to having socio-demographic impacts (Montgomery et al., 2013), also affects the environment and the economy (Dameri et al., 2019; Cohen, 2015; Nam & Pardo, 2011a, 2011b). And it is in response to these issues that the concept of the smart city was born. To understand how this concept can be considered a solution to the sociodemographic, environmental, and economic challenges posed by increasing urbanization, it becomes necessary to define and analyze it. The term "smart city" is first used in the 1990s focusing "on the significance of the new ICT with regard to modern infrastructures within cities" (Albino et al., 2015, p. 4). Therefore, in generic terms, the smart city can be defined as "an urban environment that utilizes ICT and other related technologies to enhance performance efficiency of regular city operations and quality of services (QoS) provided to urban citizens" (Silva et al., 2018, p. 697). However, this term is a difficult concept to define. In fact, there is no universally accepted definition (Laufs et al., 2020). Ramaprasad, Sánchez-Ortiz, and Syn identified as many as 36 different definitions of smart city, but nevertheless they could not find one that encapsulates them all within it (Ramaprasad et al., 2017). For this reason, the smart city concept is defined as "fuzzy," meaning vague and difficult to identify (Angelidou, 2014; Caragliu et al., 2009; Chourabi et al., 2012; Giffinger et al., 2007; Hollands, 2008; Meijer & Bolívar, 2016; Nam & Pardo, 2011a; Neirotti et al., 2014; Nesti, 2018). This is due to the fact that studies conducted on smart cities belong to different disciplinary areas and with therefore completely different research cuts between them (Nesti, 2018). However, the most important definitions can be schematized as follows (Table 1.1).

As can be seen from Table 1.1, the smart city concept is quite popular in recent years. Most research focuses on the role of ICT infrastructure, although more and more research conducted on the role of human, social, and relationship capital, but especially environmental issues as key drivers of urban growth is gaining weight (Caragliu et al., 2009). As could be understood therefore, the smart city concept is a concept born from the union of different fields of research. For this reason, its characteristics vary depending on the research and the corresponding cut (Mozuriunaite & Sabaitytė, 2021). However, thanks to the increasing interest of scholars in this concept, it has come more and more over the years to delineate its boundaries and characteristics in a uniform way (Nesti, 2018). In fact, Meijer and Bolìvar in 2015 came to identify three research focuses around which the smart city concept revolves: ICT, people, and governance. To these Nesti (2018) added the fourth dimension of sustainability (Fig. 1.2).

As can be seen from Fig. 1.2, the technology aspect refers to all those technologies and network infrastructures that can in some way improve the policies and lives of citizens. The "People" dimension, on the other hand, emphasizes the central role played by human capital within smart cities. Indeed, they attract highly educated and innovative citizens through their services, who in turn become the drivers of local innovation. "Governance," on the other hand, relates to the ability of smart cities to create partnerships both between businesses but also between businesses and the public administration, and this only increases investment by making them more competitive. This theme of collaboration is key to debunking a compartmentalized "siloed" bureaucracy by making services more accessible and more user-centered, that is, making services fit the diverse needs of individuals (Nesti, 2018).

Finally, cities have always been the places where the main environmental problems occur: just think of the pollution caused by means of transportation or that caused by poor waste management. And it is precisely to mitigate ongoing climate change that smart cities are studying alternative energy tools. And it is from this characteristic that the concept of "smart sustainable city" (SSC) was born. However, to understand this concept it becomes necessary to take a step back and analyze another concept at the center of public debate: sustainable development. The latter can be broken down into five historical stages as can be seen in Fig. 1.3.

It, defined in 1987 as "Sustainable development is development that meets the needs of the present without compromising the ability of future generations to meet their own needs" (WCED, 1987), was born in 1972 through two events: the Stockholm Conference and the release of the book "The Limits to Growth" by the Club of Rome (Meadows et al., 1972). These two events were followed by the Earth Summit 20 years later, which resulted in the resolution of three key environmental

Table 1.1 Smart city definitions (Source: Albino et al., 2015; Cocchia, 2014)

Reference	Definition
Bakıcı et al. (2013)	Smart city as a high-tech intensive and advanced city that connects people, information and city elements using new technologies in order to create a sustainable, greener city, competitive and innovative commerce, and an increased life quality
Barrionuevo et al. (2012)	Being a smart city means using all available technology and resources in an intelligent and coordinated manner to develop urban centers that are at once integrated, habitable, and sustainable
Caragliu et al. (2009)	A city is smart when investments in human and social capital and traditional (transport) and modern (ICT) communication infrastructure fuel sustainable economic growth and a high quality of life, with a wise management of natural resources, through participatory governance
Chen (2010)	Smart cities take advantage of communications and sensor capabilities sewn into the cities' infrastructures to optimize electrical, transportation, and other logistical operations supporting daily life, thereby improving the quality of life for everyone
Cretu (2012)	Two main streams of research ideas: (1) smart cities should do everything related to governance and economy using new thinking paradigms and (2) smart cities are all about networks of sensors, smart devices, real-time data, and ICT integration in every aspect of human life
Dameri (2013)	A smart city is a well-defined geographical area, in which high technologies such as ICT, logistic, energy production, and so on, cooperate to create benefits for citizens in terms of Well-being, inclusion and participation, environmental quality, intelligent development; it is governed by a well-defined pool of subjects, able to state the rules and policy for the city government and development
Eger (2009)	Smart community—a community which makes a conscious decision to aggressively deploy technology as a catalyst to solving its social and business needs—Will undoubtedly focus on building its high-speed broadband infrastructures, but the real opportunity is in rebuilding and renewing a sense of place, and in the process a sense of civic pride. [...] smart communities are not, at their core, exercises in the deployment and use of technology, but in the promotion of economic development, job growth, and an increased quality of life. In other words, technological propagation of smart communities isn't an end in itself, but only a means to reinventing cities for a new economy and society with clear and compelling community benefit
Giffinger et al. (2007)	A city well performing in a forward-looking way in economy, people, governance, mobility, environment, and living, built on the smart combination of endowments and activities of self-decisive, independent and aware citizens. Smart city generally refers to the search and identification of intelligent solutions which allow modern cities to enhance the quality of the services provided to citizens
Guan (2012)	A smart city, according to ICLEI, is a city that is prepared to provide conditions for a healthy and happy community under the challenging conditions that global, environmental, economic and social trends may bring
Hall et al. (2000)	A city that monitors and integrates conditions of all of its critical infrastructures, including roads, bridges, tunnels, rails, subways, airports, seaports, communications, water, power, even major buildings, can better optimize its resources, plan its preventive maintenance activities, and monitor security aspects while maximizing services to its citizens

(continued)

Table 1.1 (continued)

Reference	Definition
Harrison et al. (2010)	A city connecting the physical infrastructure, the IT infrastructure, the social infrastructure, and the business infrastructure to leverage the collective intelligence of the city
Komninos (2011)	(smart) cities as territories with high capacity for learning and innovation, which is built-in the creativity of their population, their institutions of knowledge creation, and their digital infrastructure for communication and knowledge management
Kourtit and Nijkamp (2012)	Smart cities are the result of knowledge-intensive and creative strategies aiming at enhancing the socio-economic, ecological, logistic and competitive performance of cities. Such smart cities are based on a promising mix of human capital (e.g. skilled labor force), infrastructural capital (e.g. high-tech communication facilities), social capital (e.g. intense and open network linkages) and entrepreneurial capital (e.g. creative and risk-taking business activities)
Kourtit et al. (2012)	Smart cities have high productivity as they have a relatively high share of highly educated people, knowledge-intensive jobs, output-oriented planning systems, creative activities and sustainability-oriented initiatives
Lazaroiu and Roscia (2012)	A community of average technology size, interconnected and sustainable, comfortable, attractive and secure
Lombardi et al. (2012)	The application of information and communications technology (ICT) with their effects on human capital/education, social and relational capital, and environmental issues is often indicated by the notion of smart city
Nam and Pardo (2011a)	A smart city infuses information into its physical infrastructure to improve conveniences, facilitate mobility, add efficiencies, conserve energy, improve the quality of air and water, identify problems and fix them quickly, recover rapidly from disasters, collect data to make better decisions, deploy resources effectively, and share data to enable collaboration across entities and domains
Su et al. (2011)	Smart City is the product of Digital City combined with the internet of things
Thite (2011)	Creative or smart city experiments [...] aimed at nurturing a creative economy through investment in quality of life which in turn attracts knowledge workers to live and work in smart cities. The nexus of competitive advantage has [...] shifted to those regions that can generate, retain, and attract the best talent
Thuzar (2011)	Smart cities of the future will need sustainable urban development policies where all residents, including the poor, can live well and the attraction of the towns and cities is preserved. [...] smart cities are cities that have a high quality of life; those that pursue sustainable economic development through investments in human and social capital, and traditional and modern communications infrastructure (transport and information communication technology); and manage natural resources through participatory policies. Smart cities should also be sustainable, converging economic, social, and environmental goals
Washburn and Sindhu (2010)	The use of smart computing technologies to make the critical infrastructure components and services of a city—Which include city administration, education, healthcare, public safety, real estate, transportation, and utilities—More intelligent, interconnected, and efficient

(continued)

Table 1.1 (continued)

Reference	Definition
Zygiaris (2013)	A smart city is understood as a certain intellectual ability that addresses several innovative socio-technical and socio-economic aspects of growth. These aspects lead to smart city conceptions as "green" referring to urban infrastructure for environment protection and reduction of CO2 emission, "interconnected" related to revolution of broadband economy, "intelligent" declaring the capacity to produce added value information from the processing of city's real-time data from sensors and activators, whereas the terms "innovating", "knowledge" cities interchangeably refer to the city's ability to raise innovation based on knowledgeable and creative human capital

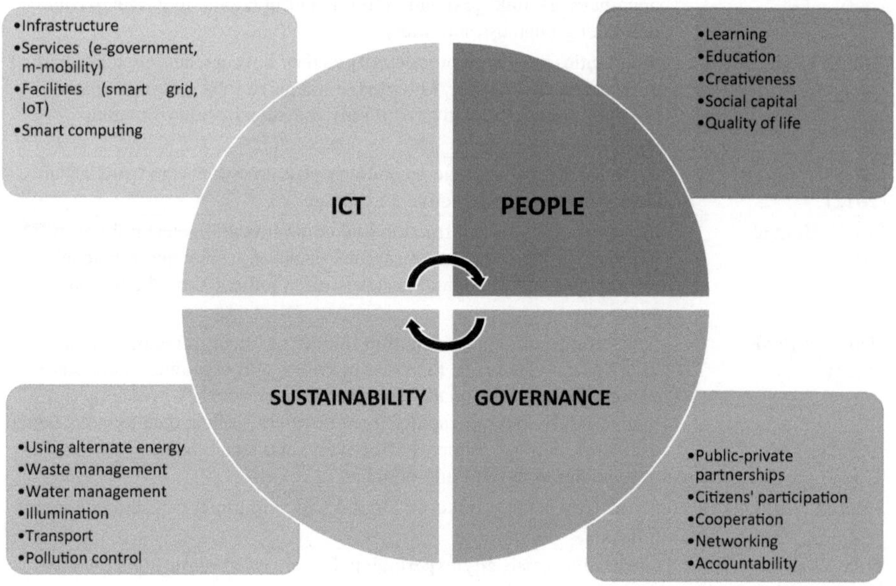

Fig. 1.2 Smart city dimensions (Source: Nesti, 2018)

documents: the Rio Declaration on Environment and Development, the Principles on Forests, and Agenda 21 (Tokuç, 2013). In particular, the latter outlined 27 principles to be achieved during the twenty-first century. Eight years later in its wake, the United Nations Millennium Declaration was signed, in which eight goals, the Millennium Development Goals (MDGs) to be achieved by 2015, were defined. And in the very year the MDGs expired, the famous and current 2030 Agenda was adopted in which 17 goals—the so-called Sustainable Development Goals (SDGs)—and 169 subgoals (targets) were defined.

These objectives eventually gave rise to the European Green Deal, which was introduced by the European Union and offers an action plan that promotes resource

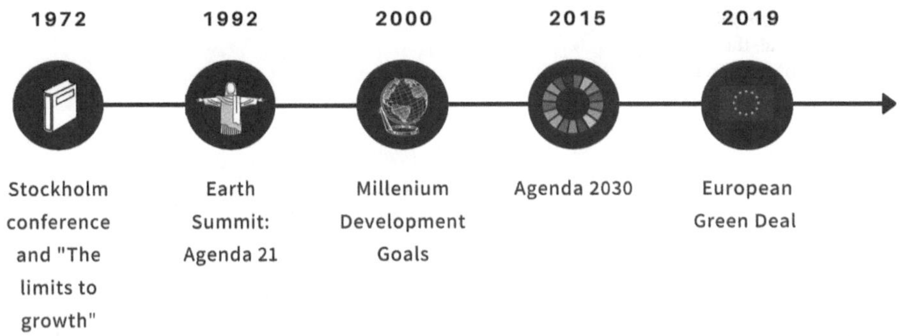

Fig. 1.3 Historical stages of sustainable development

efficiency, a clean, circular economy, biodiversity restoration, and pollution reduction. In fact, achieving carbon neutrality by 2050 is the ultimate European ambition (Toni, 2015; Rifkin, 2019). Therefore, sustainable development refers to economic and social advancement that is compatible with preserving the environment, promoting social justice, and upholding the rights of future generations. The Bruntland Report's direction has the advantage of making clear the areas on which national and international strategies should concentrate. In particular, it brought attention to the interdependence of social, economic, and environmental development by advocating a comprehensive strategy that places a focus on environmental sustainability. The Report also presents the concept of long-term thinking, the criteria of equality and justice between and between generations, and, lastly, the idea of efficient resource utilization. The rule of the balance of the three E's—economic, equity, and environment—was developed using these ideas (WCED, 1987). Economic means having the capacity to raise GDP per capita and boost competitiveness without reducing wages and employment. Equity is the improvement in a person's security, health, nutrition, education, and respect for their basic rights. The capacity to produce resources, absorb harmful externalities, and create benefit is what is referred to as environment.

A Venn diagram (Fig. 1.4) that shows how these components interact shows how the three circles, each of which represents a different concept, come together to form sustainable development as a whole. In truth, the three pillars of sustainable development have steadily grown more complicated, leading to the present 17 Sustainable Development Goals, as the modern world has grown more complex (SDGs). The 2030 Agenda has also brought about the political aspect of this, which is the requirement to reach global consensus for new governance to enable shared and coordinated transformation.

Because of this, the UNESCO introduced a fourth pillar to its structure in 2001: the preservation of cultural diversity. Harmonization across these four pillars proves to be highly challenging since it calls for coordination, but more importantly, reconciliation of topics and interests that usually clash. Equity between and among generations is yet another key idea in sustainability. The latter is characterized as

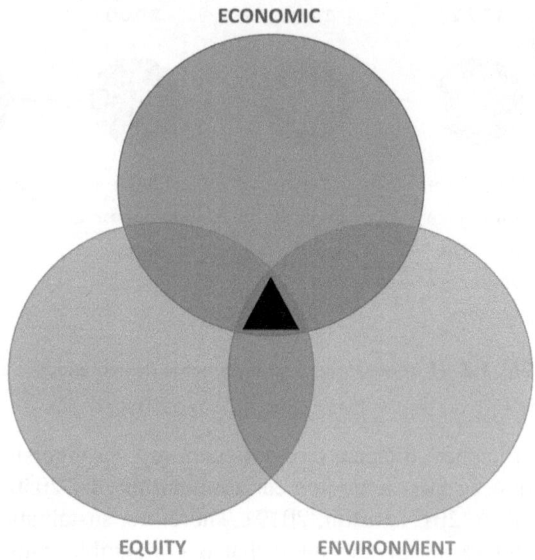

Fig. 1.4 Sustainable development as the interaction of the three E's

present generations' moral obligation to provide future generations with equal possibilities for advancement. Therefore, each generation has a responsibility to protect biodiversity, maintain the condition of the world, and guarantee access to an abundance of natural and cultural resources in a way that ensures future generations will have the opportunity to make their own decisions. This idea has dual value since it (a) ensures equality between the more developed and less developed nations and (b) allows for the integration of minorities (Giovannini, 2018).

1.2 From the SSC Concept to the Smart Sustainable Knowledge City Concept

Therefore, SSCs indicate cities where the quality of life, livability, technological development, and natural environment protection should come first in the eyes of policymakers. All these factors together contribute to attracting and retaining knowledge-based workers and companies. In other words, the interaction between the SSC pillars generates agglomeration economies (Penco, 2015). People and, therefore, workers look at the quality of life factors to establish their location preferences. We recognize strategic and operational factors of attraction (Carrillo, 2004; Ergazakis et al., 2004; Yigitcanlar, 2011). The strategic factors of attraction are the educational system, financial incentives, public policies, and urban planning. The operational factors are represented by the Internet connection network, transport system, R&D centers, entrepreneurial clusters, and, more in general, all the built-in assets that facilitate the localization of knowledge workers and enterprises. Both strategic and operational factors contribute to urban development (Penco, 2015). In addition, SSC's attraction potential directly depends on the city's capacity to support its inhabitants' leisure time. For this reason, amenities, cinemas, museums, theaters, art galleries, green spaces, and entertainment operators facilitate the attraction of people and workers by enhancing the quality of life (Edvinsson, 2006; Glaeser & Gottlieb, 2006; Yigitcanlar et al., 2007).

Another element that contributes to attracting knowledge workers is the "people climate" (Florida, 2002). The people's climate, in the name of diversity, is a relevant factor of attraction for knowledge workers. The people's climate identifies the specific behavioral traits of the city and confers a precise identity to the city itself. The ideal SSC identity is those of a tolerant and open to new ideas city based on the triple pillar of technology-talent-tolerance (Florida, 2002).

Attracting knowledge workers means developing the creative engine of a city. This trend benefits entrepreneurship since an enhanced knowledge potential creates positive spillovers in terms of a capacity of an area to undertake entrepreneurial activities, commercialize innovative products, and create synergies between centers of knowledge production (e.g., universities and companies). Specifically, the triple helix formed by knowledge-intensive firms, education and research institutions, and knowledge-intensive services (e.g., financial services, telecommunication networks, and international public institutions) generates closed transmission of knowledge (Ivaldi et al., 2020).

1.3 The Mutual Relationships between SSC and Knowledge-Based Economy

Therefore, the SSC characteristics strengthen the premises for a knowledge city can prosper. A knowledge city is defined as a city whose purpose is to anchor to and develop a knowledge economy (Carrillo, 2004) through an enhanced service orientation of their economic activities (Glaeser et al., 2010, 2014; Sassen, 1994; Turok, 2008). This knowledge "production system" is grounded on inbound and outbound knowledge exchanges involving actors internal to the city and agents operating in other cities. These exchanges are enabled by knowledge management processes powered by ICT networks and infrastructures (Ergazakis et al., 2004; Pancholi et al., 2019).

However, SSC and the knowledge-based economy are characterized by a bijective relationship. The principle is that mutually reinforcing mechanisms work in order for the SSC environment strengthens the knowledge-based economy and, at the same time, the knowledge-based economy dynamism improves SSC. In fact, SSC contributes to the growth of the knowledge-based economy by providing facilitations for people, knowledge workers, and entrepreneurs. In parallel, these crucial actors develop practical solutions to address the needs of the urban community. This mutual exchange materializes, for instance, by increasing the knowledge capital of the city and strengthening the networks between entrepreneurs.

Ivaldi et al. (2020) studied the interrelationship between SSC and knowledge-based economy through a correlation analysis based on two indexes, i.e., the SSC index and the knowledge-economy index. Overall, the results showed a moderate correlation between the two indexes confirming the relevant role of an SSC toward knowledge-based economic development. The analysis reveals a higher correlation between the SSC's dimensions of services and local transport and the knowledge-based economy. This finding confirms the service-oriented nature of the knowledge economy and the relevance of the transportation system as an operational factor.

Concerning urban size, Ivaldi et al. (2020) showed that larger cities are more prone to improve knowledge-based economies and innovations. In parallel, small cities are more environmentally sustainable because they are characterized by better green spaces, higher quality of air, and less noise pollution. Therefore, the development of a smart and sustainable knowledge identity is also possible in small cities. Large metropolitan cities leverage agglomeration economies, positive externalities, spillovers, economies of scale, innovation, and growth to fuel their smart and sustainable knowledge traits (Chatterji et al., 2014; Szerb et al., 2013). Small cities attract creative classes by enhancing the quality of life (Ström & Nelson, 2010). In these terms, small cities should be forefront cities in promoting the community's needs and livability issues (Waitt & Gibson, 2009). Therefore, both large and small cities can achieve a knowledge-based economy SSC configuration through two different strategic approaches.

The authors wrote that, at the same time, "the development of a knowledge-based economy can improve the level of smartness and sustainability of the city since the knowledge workers express a specific demand for city living and for high quality of life" (Florida et al., 2008; Yigitcanlar et al., 2007). Therefore, the work by Ivaldi et al. (2020) represented one step forward in studying the mutually influencing relationships between SSC and the knowledge economy.

Zurich represents a virtuous example of SSC. It implemented a strategy to meet the future needs of the population and promote innovation and smartness increase (Zurich City Council, 2018). Overall, these measures strengthen the collaboration between politics, business, science, culture, and society. In this way, the urban entrepreneurial ecosystem can prosper fueled by high digitization levels and a propensity toward innovation. Attention toward the urban community's needs makes the city a proactive system in response to the most recent topics. Therefore, Zurich represents an example of SSC engaged in a steady effort to satisfy emerging human needs through digital-fueled and participated solutions promoted with the assistance of an innovative knowledge-based entrepreneurial fabric. Robust and advanced city's digital architecture equips Zurich with a valuable infrastructure base for businesses and services of the knowledge-based economy (Aina, 2017; Anthopoulos, 2017; Bibri & Krogstie, 2017).

Aalborg has invested many resources in sustainability issues. The SSC of Aalborg is a sustainability-driven model. In addition, the Aalborg plan for SSC involves policies to guarantee high employment rates for its citizens, areas for idea sharing, and citizen action promotion. To strengthen its knowledge economy and infrastructural fabric, Aalborg implemented holistic solutions aimed at digitizing, innovating, and building ICT architecture (Smart City Press, 2020).

Vienna showed a commitment similar to Aalborg to promote solutions against climate change. By deliberating the "Smart City Wien" strategy (2014) and subsequent government agreement (2020), Vienna fixed the target of climate neutrality by 2040. Overall, the broader mission of this strategy is to improve the quality of life for the urban community through social and technical innovations that maximize the conservation of natural resources. To address this target, Vienna aims to use social inclusion, youth policies, energy conservation, climate change fight, and innovation and digitalization as main development levers (Smart City Wien, 2022).

Cardiff defined a "Smart City Roadmap" (2021) to drive its SSC strategies. It is a multi-step document of urban development toward a smart and sustainable knowledge city model (Cardiff Council's Smart City Road Map, 2021). The purpose is to improve public services' effectiveness and efficiency and energy infrastructure, innovate public transport to reduce travel time, and capitalize on the economic and social cascade effects generated by digitization and modernization of the urban environment. Overall, this urban development plan should increase productivity and fuel economic growth. The knowledge-based economy should benefit from enhanced levels of digitization and connectivity. At the same time, Cardiff could strengthen its role as a knowledge worker pole of attraction by offering modern services and high quality of life.

These paradigmatic examples show that successful smart and sustainable knowledge cities follow similar development patterns. Differences emerge for the valorization of urban peculiarities and resource availability. The pillars sustaining a smart and sustainable knowledge city are those of human centricity, service orientation, digitization, innovation promotion, infrastructural modernization, environmental sustainability (e.g., carbon emission reduction, low-impact public transport, renewable energy, etc.), and participative formulation of the public policies.

1.4 Conclusions

This brief background on the concept of sustainable development and knowledge-based economy is useful to fully understand the concept of smart sustainable knowledge city. For the UNECE and ITU, "A SSC is an innovative city that uses ICT and other means to improve quality of life, efficiency of operation and urban services, and competitiveness, while ensuring that it meets the needs of present and future generations in economic, social, environmental and cultural aspects (UNECE, 2022)."

The concept of SSC, thus defined, makes it clear how cities are increasingly a center of human, service-oriented economic and environmental activity. In fact, the smart sustainable knowledge city was created to respond to all those difficult-to-manage problems, thereby improving social cohesion and environmental sustainability and consequently causing greater personal satisfaction among its citizens (Keshvardoost et al., 2018; Monzon, 2015).

Cities must carefully balance their problems as they expand relentlessly to ensure that population expansion, economic development, and social progress all occur at roughly the same pace. Although metropolitan areas produce most of the world's GDP, not all activities there produce positive externalities. Cities also have the highest levels of inequality; if these problems are not effectively addressed, the negative effects can outweigh the positive ones. The smart sustainable knowledge city model can help a city become more organized and function better and, in this way, can also help a city realize a realistic model of urban expansion (Keshvardoost et al., 2018).

Therefore, smart sustainable knowledge cities can be seen as open ecosystems in which different actors (political actors, public administration, associations, research centers, universities, businesses, and citizens) collaborate with each other to generate innovation with the basic assumption that knowledge is diffused within society and therefore new solutions should be sought both internally and externally (Chesbrough, 2003). Finally, research on smart cities is still largely in progress. Local peculiarities, civic goals and needs, as well as global market forces and accessible technology influence the terrain of the smart city (Angelidou, 2014).

Kigali

Country: Rwanda

Inhabitants: 1,132,686

Density: 1552/km^2 (4020/sq. mi)

Total area: 730 km^2 (280 sq. mi)

Kigali is the capital of the landlocked country of Rwanda, with a population of over 13 million people in Central Africa. The city government aspires to establish itself as a scalable model within its continent and serve as a test case for other African countries by pursuing social inclusion, economic growth, and sustainable development. Kigali has set a number of significant objectives that must be met in order to become a Smart City, such as making city infrastructures and the managing of urban resources sustainable, lowering housing costs to be affordable to everyone, conserving green spaces, and preserving citizens' health by fostering light mobility.

The establishment of district-specific plans for the numerous neighborhoods in the capital, as well as a general master plan for the entire city of Kigali, are important elements of the methodical action plan that was created to achieve the aforementioned goals. These programs outline the rules for monitoring and coordinating activities. These actions include building the transportation infrastructure required to redesign city mobility, enhancing the system for collecting taxes and moving commodities within cities, and attracting foreign capital.

(continued)

Building new connections, providing for total lighting of urban spaces, and enhancing management of liquid and gaseous waste are all goals of a comprehensive restructuring program for the existing network that is required for Kigali to become a Smart City. The municipal government's task is to formulate laws to encourage investment by making it easier for foreign capital to enter the city. To that purpose, corporate tax incentive programs have been promoted, and a platform connecting businesses with local government has been developed.

Kigali has promoted social support initiatives for vulnerable groups, including children, the unemployed, the elderly, and the disabled. To keep the city's streets and urban areas clean, the local administration works with associations and cooperatives that have grown over time. The capital municipality of Rwanda is actively seeking to emancipate women because it recognizes the importance of gender equality on both a social and economic level. Education is also a cornerstone of Kigali's Smart City strategy: The introduction of academic awards has made student rivalry more compelling and has attempted to raise the level of culture generally.

In conclusion, Kigali is one of the few African cities that have shown a commitment to developing into a Smart City in the near future: Once the needs have been identified and the goals have been established, the implementation phase is now underway.

References

Albino, V., Berardi, U., & Dangelico, R. M. (2015). Smart cities: Definitions, dimensions, performance, and initiatives. *Journal of Urban Technology, 22*(1), 3–21. https://doi.org/10.1080/10630732.2014.942092

Aina, Y. A. (2017). Achieving smart sustainable cities with GeoICT support: The Saudi evolving smart cities. *Cities, 71*, 49–58.

Angelidou, M. (2014). Smart city policies: A spatial approach. *Cities, 41*, S3–S11. https://doi.org/10.1016/j.cities.2014.06.007

Anthopoulos, L. (2017). Smart utopia VS smart reality: Learning by experience from 10 smart city cases. *Cities, 63*, 128–148.

Bakıcı, T., Almirall, E., & Wareham, J. (2013). A Smart City initiative: The case of Barcelona. *Journal of the Knowledge Economy, 4*(2), 135–148.

Barrionuevo, J., Berrone, P., & Ricart, J. (2012). Smart cities, sustainable progress: Opportunities for urban development. *IESE Insight*, 50–57. https://doi.org/10.15581/002.ART-2152

Bibri, S. E., & Krogstie, J. (2017). Smart sustainable cities of the future: An extensive interdisciplinary literature review. *Sustainable Cities and Society, 31*, 183–212.

Caragliu, A., Bo, C. D. D., & Nijkamp, P. (2009). *Smart Cities in Europe*. https://doi.org/10.1080/10630732.2011.601117

Cardiff Council's Smart City Road Map. (2021). *Smart Cardiff.* 0000765 RC Smart Cities 2019 FIN craig.indd. Accessed Oct 23, 2022, from (smartcardiff.co.uk)

Carrillo, F. J. (2004). Capital cities: A taxonomy of capital accounts for knowledge cities. *Journal of Knowledge Management, 8*(5), 28–46.

Chatterji, A., Glaeser, E., & Kerr, W. (2014). Clusters of entrepreneurship and innovation. *Innovation Policy and the Economy, 14*(1), 129–166.

Chen, T. M. (2010). Smart grids, smart cities need better networks [Editor's Note]. *IEEE Network, 24*(2), 2–3. https://doi.org/10.1109/MNET.2010.5430136

Chesbrough, H. (2003). *Open innovation: The new imperative for creating and profiting, from technology*. Harvard Business School Press.

Chourabi, H., Nam, T., Walker, S., Gil-Garcia, J. R., Mellouli, S., Nahon, K., Pardo, T., & Scholl, H. (2012). Understanding smart cities: An integrative framework. In *45th Hawaii International Conference on System Sciences* (pp. 2289–2297). https://doi.org/10.1109/HICSS.2012.615

Ciacci, A., Ivaldi, E., & González-Relaño, R. (2021). A partially non-compensatory method to measure the smart and sustainable level of Italian municipalities. *Sustainability, 13*(1), 435. https://doi.org/10.3390/su13010435

Cocchia, A. (2014). Smart and digital city: A systematic literature review. In I. R. P. Dameri & C. Rosenthal-Sabroux (Eds.), (A c. Di) *Smart city* (pp. 13–43). Springer International Publishing. https://doi.org/10.1007/978-3-319-06160-3_2

Cohen, B. (2015). *Urbanization, city growth, and the New United Nations development agenda*. 72.

Cretu, L. G. (2012). Smart cities design using event-driven paradigm and semantic web. *Informatica Economica, 16*(4), 57–67.

Dameri, R. (2013). Searching for smart city definition: A comprehensive proposal. *International Journal of Computers and Technology, 11*, 2544. https://doi.org/10.24297/ijct.v11i5.1142

Dameri, R. P., Benevolo, C., Veglianti, E., & Li, Y. (2019). Understanding smart cities as a glocal strategy: A comparison between Italy and China. *Technological Forecasting and Social Change, 142*, 26–41. https://doi.org/10.1016/j.techfore.2018.07.025

Edvinsson, L. (2006). Aspects on the city as a knowledge tool. *Journal of Knowledge Management, 10*(5), 6–13.

Eger, J. M. (2009). Smart growth, smart cities, and the crisis at the pump A worldwide phenomenon. *I-WAYS, Digest of Electronic Commerce Policy and Regulation, 32*(1), 47–53. https://doi.org/10.3233/IWA-2009-0164

Eremia, M., Toma, L., & Sanduleac, M. (2017). The smart city concept in the 21st century. *Procedia Engineering, 181*, 12–19. https://doi.org/10.1016/j.proeng.2017.02.357

Ergazakis, K., Metaxiotis, K., & Psarras, J. (2004). Towards knowledge cities: Conceptual analysis and success stories. *Journal of Knowledge Management, 8*(5), 5–15.

Florida, R. (2002). The economic geography of talent. *Annals of the Association of American Geographers, 92*(4), 743–755.

Florida, R., Mellander, C., & Stolarick, K. (2008). Inside the black box of regional development—Human capital, the creative class and tolerance. *Journal of Economic Geography, 8*(5), 615–649.

Giffinger, R., Fertner, C., Kramar, H., Kalasek, R., Milanović, N., & Meijers, E. (2007). *Smart cities—Ranking of European medium-sized cities*.

Giovannini, E. (2018). *L'utopia sostenibile* (Italian ed.). Editori Laterza.

Glaeser, E. L., & Gottlieb, J. D. (2006). Urban resurgence and the consumer city. *Urban Studies, 43*(8), 1275–1299.

Glaeser, E. L., Ponzetto, G. A., & Tobio, K. (2014). Cities, skills and regional change. *Regional Studies, 48*(1), 7–43.

Glaeser, E. L., Rosenthal, S. S., & Strange, W. C. (2010). Urban economics and entrepreneurship. *Journal of Urban Economics, 67*(1), 1–14.

Guan, L. (2012). Smart steps to a better city. *Government News, 32*(2), 24–27. https://doi.org/10.3316/informit.521507841779512

Hall, R., Bowerman, B., Braverman, J., Taylor, J., Todosow, H., and Wimmersperg, U. (2000). *The vision of a smart city*. 2nd Int Life.

Harrison, C., Eckman, B., Hamilton, R., Hartswick, P., Kalagnanam, J., Paraszczak, J., & Williams, P. (2010). Foundations for smarter cities. *IBM Journal of Research and Development, 54*, 1. https://doi.org/10.1147/JRD.2010.2048257

Hoballah, A., Peter, C., & Programme des Nations Unies pour l'environnement. (2012). *Sustainable, resource efficient cities: Making it happen!*

Hollands, R. (2008). Will the real smart city please stand up? *City, 12*, 303–320. https://doi.org/10.1080/13604810802479126

Ivaldi, E., Penco, L., Isola, G., & Musso, E. (2020). Smart sustainable cities and the urban knowledge-based economy: A NUTS3 level analysis. *Social Indicators Research, 150*(1), 45–72.

Keshvardoost, S., Renukappa, S., & Suresh, S. (2018). Developments of policies related to smart cities: A critical review. In *2018 IEEE/ACM International Conference on Utility and Cloud Computing Companion (UCC Companion)* (pp. 370–375). https://doi.org/10.1109/UCC-Companion.2018.00083

Komninos, N. (2011). Intelligent cities: Variable geometries of spatial intelligence. *Intelligent Buildings International, 3*, 172–188. https://doi.org/10.1080/17508975.2011.579339

Kourtit, K., & Nijkamp, P. (2012). Smart cities in the innovation age. *Innovation: The European Journal of Social Science Research, 25*(2), 93–95. https://doi.org/10.1080/13511610.2012.660331

Kourtit, K., Nijkamp, P., & Arribas-Bel, D. (2012). Smart cities in perspective–A comparative European study by means of self-organizing maps. *Innovations, 25*(2), 229–246. https://doi.org/10.1080/13511610.2012.660330

Laufs, J., Borrion, H., & Bradford, B. (2020). Security and the smart city: A systematic review. *Sustainable Cities and Society, 55*, 102023. https://doi.org/10.1016/j.scs.2020.102023

Lazaroiu, G. C., & Roscia, M. (2012). Definition methodology for the smart cities model. *Energy, 47*(1), 326–332. https://doi.org/10.1016/j.energy.2012.09.028

Lombardi, P., Giordano, S., Farouh, H., & Yousef, W. (2012). Modelling the smart city performance. *Innovation: The European Journal of Social Science Research, 25*(2), 137–149. https://doi.org/10.1080/13511610.2012.660325

Meadows, D. H., Meadows, D. L., Randers, J., & Behrens, W. W., III. (1972). *The limits to growth* (p. 1972). Universe Books.

Meijer, A., & Bolívar, M. (2016). *Governing the smart city: A review of the literature on smart urban governance*. https://doi.org/10.1177/0020852314564308.

Montgomery, M., Stren, R., Cohen, B., & Reed, H. (2013). *Cities transformed: Demographic change and its implications in the developing world*. 9781315065700, 1–533. https://doi.org/10.4324/9781315065700.

Monzon, A. (2015). Smart cities concept and challenges: Bases for the assessment of smart city projects. In *Smart cities and green ICT systems (SMARTGREENS), 2015 International Conference on* (pp. 1–11). SCITEPRESS.

Mozuriunaite, S., & Sabaitytė, J. (2021). To what extent we do understand smart cities and characteristics Influencing City smartness. *Journal of Architecture and Urbanism, 45*, 1–8. https://doi.org/10.3846/jau.2021.12392

Nam, T., & Pardo, T. (2011a). *Conceptualizing smart city with dimensions of technology, people, and institutions* (pp. 282–291). https://doi.org/10.1145/2037556.2037602

Nam, T., & Pardo, T. A. (2011b). Smart city as urban innovation: Focusing on management, policy, and context. In *Proceedings of the 5th International Conference on Theory and Practice of Electronic Governance–ICEGOV'11* (p. 185). https://doi.org/10.1145/2072069.2072100

Neirotti, P., De Marco, A., Cagliano, A. C., Mangano, G., & Scorrano, F. (2014). Current trends in Smart City initiatives: Some stylised facts. *Cities, 38*, 25–36. https://doi.org/10.1016/j.cities.2013.12.010

Nesti, G. (2018). *Trasformazioni urbane. Le città intelligenti tra sfide e opportunità*. Carocci.

Pancholi, S., Yigitcanlar, T., & Guaralda, M. (2019). Place making for innovation and knowledge-intensive activities: The Australian experience. *Technological Forecasting and Social Change, 146*, 616–625.

Penco, L. (2015). The development of the successful city in the knowledge economy: Toward the dual role of consumer hub and knowledge hub. *Journal of the Knowledge Economy, 6*(4), 818–837.

Ramaprasad, A., Sánchez-Ortiz, A., and Syn, T. (2017). A unified definition of a smart city. M. Janssen, K. Axelsson, O. Glassey, B. Klievink, R. Krimmer, I. Lindgren, P. Parycek, H. J. Scholl, and D. Trutnev. (A c. Di), Electronic government (13–24). Springer International Publishing. https://doi.org/10.1007/978-3-319-64677-0_2.

Rifkin, J. (2019). *Un green deal globale*. Mondadori.

Sassen, S. (1994). The urban complex in a world economy. *International Social Science Journal, 139*, 43–62.

Silva, B. N., Khan, M., & Han, K. (2018). Towards sustainable smart cities: A review of trends, architectures, components, and open challenges in smart cities. *Sustainable Cities and Society, 38*, 697–713. https://doi.org/10.1016/j.scs.2018.01.053

Smart City Press. (2020, April 21). *8 Smallest cities in the world with the soul of A. Smart city*. Accessed Oct 23, 2022, from https://smartcity.press/small-cities-becoming-smart-cities/

Smart City Wien (2022). *Smart climate city strategy*. Accessed Oct 23, 2022, from https://smartcity.wien.gv.at/en/strategy/#top

Ström, P., & Nelson, R. (2010). Dynamic regional competitiveness in the creative economy: Can peripheral communities have a place? *The Service Industries Journal, 30*(4), 497–511.

Su, K., Li, J., & Fu, H. (2011). *Smart city and the applications*. 2011 International Conference on Electronics, Communications and Control (ICECC). https://doi.org/10.1109/ICECC.2011.6066743.

Szerb, L. A., Ács, Z., & Autio, E. (2013). Entrepreneurship and policy: The national system of entrepreneurship in the European Union and in its member countries. *Entrepreneurship Research Journal, 3*(1), 9–34.

Thite, M. (2011). Smart cities: Implications of urban planning for human resource development. *Human Resource Development International, 14*(5), 623–631. https://doi.org/10.1080/13678868.2011.618349

Thuzar, M. (2011). Urbanization in Southeast Asia: Developing smart cities for the future? In *Regional outlook* (pp. 96–100). https://doi.org/10.1355/9789814311694-022

Tokuç, A. (2013). Earth summit. In S. O. Idowu, N. Capaldi, L. Zu, & A. D. Gupta (Eds.), *Encyclopedia of corporate social responsibility*. Springer. https://doi.org/10.1007/978-3-642-28036-8_1

Toli, A. M., & Murtagh, N. (2020). The concept of sustainability in Smart City definitions. *Frontiers in Built Environment, 0*. https://doi.org/10.3389/fbuil.2020.00077

Toni, F. (2015). *I fondamenti dell'economia circolare*. Fondazione per lo Sviluppo Sostenibile.

Turok, I. (2008). A new policy for Britain's cities: Choices, challenges, contradictions. *Local Economy, 23*(2), 149–166.

UNECE. (2022). *Sustainable smart cities | UNECE. Homepage | UNECE*. https://unece.org/housing/sustainable-smart-cities

United Nations (A c. Di). (2014). *World urbanization prospects, the 2014 revision: Highlights*. United Nations.

United Nations, Department of Economic and Social Affairs, and Population Division. (2019). *World urbanization prospects: The 2018 revision*. https://www.un.org/development/desa/pd/sites/www.un.org.development.desa.pd/files/files/documents/2020/Jan/un_2018_wup_report.pdf

Yigitcanlar, T. (2011). Position paper: Redefining knowledge-based urban development. *International Journal of Knowledge Based Development, 2*(4), 340–356.

Yigitcanlar, T., Baum, S., & Horton, S. (2007). Attracting and retaining knowledge workers in knowledge cities. *Journal of Knowledge Management, 11*(5), 6–17.

Waitt, G., & Gibson, C. (2009). Creative small cities: Rethinking the creative economy in place. *Urban Studies, 46*(5–6), 1223–1246.

Washburn, D., & Sindhu, U. (2010). *Helping CIOs understand "Smart City" initiatives. Smart City*, 17.

WCED. (1987). *Our Common Future*. Oxford University Press.

Zygiaris, S. (2013). Smart City reference model: Assisting planners to conceptualize the building of Smart City innovation ecosystems. *Journal of the Knowledge Economy, 4*(2), 217–231. https://doi.org/10.1007/s13132-012-0089-4

Zurich City Council. (2018). *Strategy SMART CITY ZURICH.* Accessed Oct 23, 2022, from Smart_City_Zurich_Strategy.pdf

Chapter 2
Smart Sustainable Cities and Knowledge-Based Economy for People, Workers, and Enterprises: Mutually Reinforcing Dynamics

Andrea Ciacci and Enrico Ivaldi

Abstract The authors focus on the concepts of the knowledge-based economy and smart sustainable city (SSC) to define the nature of their interrelationships. The authors start exploring the foundations of the knowledge spillover theories. Therefore, the authors identify the complementarities between the knowledge economy and the SSC. It emerges that the SSC is a functional urban pole to accumulate more knowledge and attract creative people and knowledge workers. This tendency culminates with a mutual self-reinforcing process where knowledge spillovers and clusters of firms synergistically contribute to creating value for the city. In this chapter, the authors define the assumptions to achieve a higher level of life quality in the city and, in particular, for workers. Knowledge spillovers shape the city's structural, industrial, and social environment and influence urban development. This&spi2;chapter contributes to the literature by theorizing the symbiotic nature of the knowledge economy and SSC.

Keywords Knowledge cities · Knowledge spillovers · Urban development

2.1 Introduction

Many theories emerged to explain how new knowledge benefits an economic system. One of the most relevant is the knowledge spillover theory of entrepreneurship (KSTE) (Acs et al., 2009; Audretsch & Belitski, 2013; Qian & Acs, 2013; Tavassoli et al., 2021). However, previous literature does not capture the interrelationship between knowledge creation and SSCs. The issue is relevant if we consider that knowledge creation follows well-defined patterns influenced by an area configuration (e.g., cluster, industry systems, infrastructure availability, etc.) (Thite, 2011). By definition, SSCs are built on advanced technological systems, innovation laboratories fueled by creative and highly skilled workers, natural environment sustainability, and high quality of life standards (Ahvenniemi et al., 2017; Batty et al., 2012). Therefore, they seem to have all the qualities to stimulate knowledge

E. Ivaldi, A. Ciacci, *Smart Sustainable Cities and Knowledge-Based Economy*,
Human Well-Being Research and Policy Making,
https://doi.org/10.1007/978-3-031-25038-5_2

spillovers and, consequently, economic growth, well-being, and quality of life at a city level. This chapter aims to unveil how SSCs attract knowledge workers and stimulate the knowledge-based economy.

Therefore, this chapter focuses on the capacity of a SSC, intended as an urban environment aimed at increasing the quality of life for citizens and ensuring sustainability (Ivaldi et al., 2020; Thite, 2011; Zheng et al., 2020), to attract knowledge workers and improve the knowledge-based economy. The authors develop a framework to describe the most relevant factors of influence. Overall, it emerges that multilevel factors (e.g., institutional, organizational, individual, urban, and economic levels) influence a knowledge-based economy of a city. A city's strategy, design, infrastructure, amenities, and personality jointly determine the shape of the regional economic system attracting or rejecting creative people.

However, this mechanism is not one-directional. As such, this chapter identifies and explains the factors of the knowledge-based economy that affect SSCs focusing on the current literature and theories. The knowledge spillover stream of research is a pivotal piece of literature to explain the formation and proliferation of entrepreneurial phenomena at the city level (Acs et al., 2009; Audretsch & Belitski, 2013; Tavassoli et al., 2021). Geographical bounds characterize knowledge spillovers (Qian & Acs, 2013, p. 187). In other words, some places accumulate great knowledge stocks that renew and regenerate over time through new firm creation and the growth of existing ones. The self-sustaining process of a knowledge-intensive area contributes to the differentiation of areas within the same region or urban agglomeration. This mechanism gives rise to the creation of advantaged areas for the flows of knowledge produced (Qian & Acs, 2013). Entrepreneurs define their location preferences based on their perceptions of entrepreneurial opportunities and livability (e.g., working-personal life balance) (Audretsch et al., 2021; Tavassoli et al., 2021). For this reason, a healthy knowledge economy stimulates the new business formation and worker attraction and improves the city's reputation. The size of a knowledge-based economy shapes the city's configuration.

Overall, the deep interconnections between the SSC, its domains, and the knowledge-based economy emerge. A city's structural environment, personality, and knowledge-based economy are linked to each other by mutual processes of self-reinforcement. Given their intertwined fates, policymakers must jointly consider entrepreneurial, human, infrastructural, and social issues to define the urban development strategy. The asymmetrical development of one domain over the others seems not to reward.

2.2 The Foundations of the Knowledge Spillover-Related Theories

The term "knowledge-based economy" indicates an economy based on knowledge-creation intensity more than labor-intensive activities and immaterial production (e.g., knowledge and information) more than materials. Theories associated with the knowledge spillover explain the evolution of the knowledge-based economy. This school of thought has characterized the literature over the years and has expanded over time. It departs with Marshallian theories. Marshall (1920) theorized that knowledge spillovers stimulate firms to cluster in specific areas. In other words, knowledge spillover facilitates the firms' agglomeration within specific areas. Where knowledge overflows with higher intensity, firms tend to agglomerate.

The theory by Jacobs represents another milestone to understand the relevance of knowledge spillover in an urban setting (Jacobs, 1961, 1969). Jacobs enlarged the theory beyond the economic side by including factors related to individual psychology. To not misinterpret the theory by Jacobs, it is necessary to give due importance to the people living in the city and not only to the firms and industries structure of the city itself. Specifically, Jacobs highlights the city's human side by looking at contact and interaction among people (e.g., casting "eyes on the street"). However, Jacobs' conceptualization is not limited to individual and psychological traits. It consists of an approach based on two non-mutually exclusive categories. In addition to people, Jacobs and further scholars look at the structural environment of the city. This perspective embraces the idea that density, i.e., a physical characteristic of the city, and industry diversity, i.e., a city's functional characteristic, jointly support knowledge spillovers between industries, facilitate new knowledge generation, the proliferation of innovative activities and products/services, and diffusion of high-quality entrepreneurship. Even cities with the same structural environment in terms of density and diversity can obtain different results in their entrepreneurship activity and related economic performance (Agrawal et al., 2014; Roche, 2019). In fact, density and diversity are not the only characteristics of a city's structural environment to play a crucial role in the economic field. For instance, the interplay between diversity and specialization affects the responsiveness of the economy and entrepreneurial system. A dichotomy often arises between diversity and specialization as two distinguished types of knowledge spillovers. Authors have worked to clarify what condition is the best in promoting an effective knowledge spillover within city boundaries. The results diverge, showing a significant heterogeneity. However, an unambiguous conclusion highlights diversity tends to have a more robust positive influence on new knowledge generation, creativity, (product) innovation, and high-quality entrepreneurship. At the same time, specialization seems to be more conducive to efficiency and labor productivity.

In summary, individuals are characterized by personality traits deriving from psychology (Pervin & John, 1999). People differ in terms of psychological characteristics in the same way cities and regions distinguish for their personalities. In fact, the notion of personality applies both to individuals and geographical areas (Florida, 2010). For instance, Tavassoli et al. (2021) studied the city's personality as local openness. The authors found that local openness stimulates knowledge spillovers. The more open a city is, the more intense generation of knowledge spillovers occurs. It leads to positive effects in terms of economic performance and vitality of a city. The authors explained differences in personality traits between cities mentioning selective migration patterns and local social influence as shaping factors that determine inter-city differences in personality (Obschonka et al., 2018; Rentfrow &

Jokela, 2016; Rentfrow et al., 2008). One of the main findings of Tavassoli et al. (2021) presents the level of regional openness of cities as a predictor able to significantly and positively affect the quality of entrepreneurship. They also find that density and diversity have a positive direct effect on quality entrepreneurship. Overall, the three-way interaction involving openness, density, and diversity is positive and significant in explaining the quality of entrepreneurship. In addition, local openness interacts with a city's structural environment. Specifically, the high density and diversity of a city's structural environment increase the openness effect on quality entrepreneurship (Tavassoli et al., 2021).

Romer introduced the new growth theory (1990) to explain the knowledge production function (KPF). Over time, the new growth theory became a milestone and inspired further conceptual developments. It assumes that R&D generates innovation-related outcomes, new knowledge, and new firm growth (Jaffe, 1986; Klarl, 2013; Stam & Wennberg, 2009). Romer (1986) estimated knowledge spill-overs as the engine leading to new firm growth. In addition, new knowledge production shapes the economic system configuration at a regional level by attracting new innovative firms and clustering high-tech industries (Acs & Sanders, 2012; Stam & Wennberg, 2009). Subsequently, Romer expanded KPF involving institutional setting-related variables (Romer, 1990), ideas, population, and human capital to explain the growth.

Michael Porter massively contributed to shaping the boundaries of strategic management discipline. Nowadays, his contributions represent a milestone and constitute a key to interpreting the most strategic and competitive phenomena. A considerable part of the work by Porter aimed to study the influence of a location on a nation's competitive advantage from a strategic perspective (Porter, 1990). Cluster is a central concept of theorization by Porter (Porter, 1990). He defines clusters as "geographic concentrations of interconnected companies, specialized suppliers, service providers, firms in related industries, and associated institutions (e.g., universities, standards agencies, trade associations) in a particular field that compete but also cooperate." Regional clusters of related industries positively affect the growth of regional industries. The stronger the cluster is, the higher employment and patenting rates (e.g., a proxy of innovation performance of an area) the region achieves (Delgado et al., 2014). Porter highlighted that the growth of the regional industry also increases with the strength of within-region clusters and similar clusters in adjacent regions. Highly specialized knowledge, input, and institutions nestle in the regional clusters. In addition, a firm can meet local demand for combinations of product-service settling in a cluster or overcoming difficulties in acquiring or delivering resources (Porter, 1990, 1996). Clusters lower entry barriers and facilitate new business formation within the local system. The assemblage of different resources occurs in a cluster (e.g., availability of assets, skills, professionals, and competencies). Clusters also speed up the pace of innovation since new small and medium enterprises can better capitalize on their effort by leveraging an established plethora of partners. It is more likely that complementary bundles of firms generate "new combinations" of product, service, and knowledge toward innovation in a clustered system (Schumpeter, 1934). Finally, the mechanisms through which a cluster benefits the regional economy are to be investigated in the multiple types of externalities that arise within the cluster involving skills availability, input-output linkages, and knowledge spillovers (Delgado et al., 2014; Porter, 2000). Knowledge spillovers benefit a cluster and related dynamics such as employment growth (Delgado et al., 2014).

The knowledge spillover theory of entrepreneurship (KSTE) represents a relatively recent milestone in this research field (Acs et al., 2009; Audretsch et al., 2021; Qian & Acs, 2013; Tavassoli et al., 2021). KSTE says that entrepreneurs engaged in the new firm creation are relevant actors in the commercialization of new knowledge generated by research institutions, universities, or incumbent companies. More specifically, the knowledge spillover theory of entrepreneurship is based on two primary cornerstones: (1) not still commercialized knowledge represents a source of entrepreneurial opportunities; (2) entrepreneurs create new firms to appropriate the value of these opportunities. KSTE precisely establishes and argues a direct relationship between knowledge and entrepreneurship. The level of entrepreneurship in an area generates significant and marginal increments of the knowledge stock (Acs et al., 2009, p. 17).

2.3 City Characteristics and Factors of Attraction for Knowledge Workers

From the previous paragraph, we know that SSCs and knowledge depend on the city's capacity to attract knowledge workers, creative people, and entrepreneurs that can commercialize knowledge. In this regard, the question is which factors of SSCs attract relevant actors for the knowledge-based economy. In other words, it is crucial to identify what elements affect the rational choices of location establishment of these actors and shape the geography of the creative classes worldwide. The following paragraphs analyze the factors attracting relevant economic actors from a general point of view and from a multi-domain perspective.

2.3.1 General City Characteristics X-Ray: Cultural Diversity, Cultural Amenities, Structural Environment, and Firm Types

Basically, the mechanism of knowledge spillover entrepreneurship depends on the cultural diversity and knowledge that characterize cities and countries. Cultural diversity indicates amenities such as places where high-skilled workers can exchange ideas and frequent exposure to foreign cultures through tourism or multi-ethnic inhabitants (Audretsch et al., 2021). Different characteristics in terms of diversity and knowledge stock differently impact entrepreneurial outcomes from one place to another. As a rule of thumb, cities showing a combination of high knowledge and cultural diversity are appropriate ecosystems to favor the proliferation of entrepreneurial activities and ideas (Audretsch et al., 2021). Particularly, knowledge spillovers lead to new business creation (Van Wijk et al., 2008). Knowledge-intensive industries are more likely to benefit from local diversity than industries lacking in knowledge (Audretsch et al., 2010; Gambardella & Giarratana, 2010). In fact, the alignment of tacit non-easy-to-transfer knowledge and high cultural diversity generates a competitive advantage at the city level (Audretsch et al., 2021; Qian & Acs, 2013; Sobel et al., 2010). However, not all knowledge is transmissible. In fact, while the so-called codified, or explicit, knowledge (e.g., publications, patents) can be easier to transmit, tacit knowledge lies in people (e.g., creative workers) and does not spill over when interpersonal communications lack (Bateson, 1973; Nonaka, 1994; Polanyi, 1966). In fact, tacit knowledge is rooted in activities, actions, and individual commitment. It indwells in the mental models and represents the individual vision of the world (Johnson-Laird, 1983; Nonaka, 1994). A potential drawback is that firms operating in knowledge-intensive industries characterized by high cultural diversity must deal with higher levels of uncertainty (Audretsch et al., 2021; Robertson & Swan, 2003). The risk is that uncertainty reduces organizational efficiency (Robertson & Swan, 2003).

Cultural amenities capture skilled employees characterized by high absorptive capacity (Amezcua et al., 2020; Florida, 2002; Glaeser et al., 2001, 2004). It facilitates knowledge spillovers to occur. The most common cultural amenities are restaurants, theaters, and cinemas. They are drivers of higher lifestyle and life quality levels. Relevant amenities to attract new businesses are efficient public services, transport links, cargo terminals, warehousing, and, in general, infrastructures (Autio et al., 2014; Audretsch et al., 2015). Large cities, in particular, provide access to these relevant amenities. These functional amenities directly impact entrepreneurial vitality. Social and cultural amenities such as museums, sports arenas, and parks improve social quality of life and help attract skilled workers.

From the structural viewpoint, the density and industry diversity jointly facilitate between-industry knowledge spillovers and favor high-quality entrepreneurship. Specialization also plays a crucial role in enabling knowledge spillover. Specifically, the interplay between diversity and specialization stimulates the responsiveness of the economic and entrepreneurial systems. The optimal combination of diversity-specialization arises when they are either high or low. In addition, specialization seems to increase efficiency and labor productivity (Tavassoli et al., 2021).

Concerning the type of firm, entrepreneurial opportunities emerge when incumbent firms invest in new knowledge generation but do not commercialize the outcome. Entrepreneurship is a conduit for commercializing knowledge (Acs et al., 2009). For instance, entrepreneurs can start their own businesses to bring to the market non-commercialized knowledge. Small firms have great potential in the new knowledge generation process since they show a greater innovative capacity than large firms in terms of innovations per employee. It indicates that, without the presence of small businesses, an area could not enact economic development. Private organizations such as high-tech firms act to improve both their own performance and bring their technological advancement to the forefront of the economic and social system by leveraging knowledge spillover. In addition, many authors consider entrepreneurs as the conduit for knowledge spillovers. Therefore, entrepreneurship, involving both individual actors and private organizations, is a pivotal element of the knowledge-based economy in SSCs.

2.3.2 SSC's Multi-Domain Attracting Factors for Knowledge Workers

The SSC concept is not identifiable with a precise semantic solution. Different definitions of SSCs have emerged. The variety of these definitions indicates a complex object composed of multiple articulations. A SSC is "a city that is supported by a pervasive presence and massive use of advanced ICT, which, in connection with various urban domains and systems and how these intricately interrelate, enables cities to become more sustainable and to provide citizens with a better quality of life" (Bibri & Krogstie, 2017). This definition highlights ICT centrality, interconnections between various city domains, and sustainability together with the quality of life improvement as objectives. In addition, a SSC is defined as "an innovative city that uses [...] ICTs and other means to improve quality of life, efficiency of urban operation and services, and competitiveness, while ensuring that it meets the needs of present and future generations with respect to economic, social and environmental aspects" (ITU, 2014). The definition by ITU (2014) highlights the aspects associated with the economic competitiveness of SSCs. This definition implicitly defines economic actors, entrepreneurs, and workers as the architect of a SSC's structural aspect. Höjer and Wangel (2015, p. 10) conceive the SSC as "a city that meets the needs of its present inhabitants without compromising the ability for other people or future generations to meet their needs, and thus, does not exceed local or planetary environmental limitations, and where this is supported by ICT." This definition mainly concentrates on the sustainability-related aspects of SSC from an environmental and inter-generational perspective.

From these definitions, it is possible to identify the elements that most characterize SSCs. They are smartness (e.g., the central role of technology), sustainability, and liveability (e.g., quality of life). A SSC is a system where individual needs and aspirations do not undermine the natural heritage. At the same time, social and natural environment well-being go hand in hand like a symbiosis. ICT, technological infrastructures, and tools sustain this process of mutual development (Bibri & Krogstie, 2017). A SSC implies embracing planned and strategic development in a context where economic development strengthens while ensuring environmental protection, social equity, and justice (Bibri, 2015). Multidimensionality is an aspect that emerges when we talk about smart cities. Dirks and Keeling (2009) stress the importance of the organic integration of urban systems (e.g., transportation, energy, education, healthcare, buildings, physical infrastructure, food, water, and public safety) because, in an interconnected environment, no system should operate in isolation. Consequently, infusing intelligence into each subsystem of a city is not enough to create a SSC (Kanter & Litow, 2009). It is possible to identify different dimensions of a SSC.

As a smart city, SSC intelligently combines its resources to provide the best economic conditions (Petrolo et al., 2015). Some programs aim to foster economic growth and enhance the city's competitiveness in local and global markets creating new jobs and attracting a skilled workforce. For a SSC, economic growth should imply deploying a limited amount of natural resources. The initiatives of a SSC should find more innovative ways and solutions to overcome economic challenges (Alawadhi et al., 2012) and preserve the natural environment. Economic and environmental objectives should be equally central in SSCs. In 2016 world leaders at the United Nations adopted the New Urban Agenda, which suggests that the role of the urban economy is to promote and consolidate policies and strategies to increase the urban economic potential in relation to wealth, job opportunities, and economic resilience. Without this "inclusive, safe, resilient, and sustainable" economic growth, cities can collapse (Diamond, 2005; Glaeser, 2011; Newman et al., 2016). Economic performance depends on political, institutional, and legal environments. These three elements represent the core infrastructure of urban government (Globerman & Shapiro, 2002). The attractive capacity of a city depends on the total investments, which provide more opportunities for economic growth. For instance, favorable conditions to start a business attract investors. The ideal combination between investors' expectations and legal structure establishes when protection of property rights is guaranteed, contracts are enforced, and the promotion of collective action to maintain organizational infrastructure is encouraged. Only good governance permits to reach of these goals (Dixit, 2009). Due to the many connections that bring the economic sphere closer to all other spheres, framework conception is a decisive factor to achieve the best economic outcomes (Allam & Newman, 2018).

Smart governance means various stakeholders are engaged in decision-making and public services. ICT-mediated governance (e.g., e-governance) is a fundamental tool in bringing SSC initiatives to citizens and keeping the decision and implementation process transparent. Participative and inclusive approaches to decision-making empower the city's inhabitants. In addition, such an approach is only acceptable to create cohesion between multiple stakeholders. The spirit of e-governance in a smart city should be citizen-centric and citizen-driven (Giffinger et al., 2007; Thuzar, 2011). Smart governance, therefore, should facilitate the relations between institutions and citizens. A new model of governance reflecting a smarter vision has to look for more inclusive processes during the various phases of public politics. Greater involvement of the different actors operating at the different levels of decision-making and relational spheres is an indispensable condition to enable value automatisms (Dameri, 2012; Neirotti et al., 2014; Washburn et al., 2009).

The smart people factor comprises various aspects, such as social and ethnic plurality, flexibility, creativity, cosmopolitanism, open-mindedness, and participation in public life (Nam & Pardo, 2011). A smart city is human-centered and offers multiple opportunities for human capital (Winters, 2010). Creativity, human capital, and cooperation among relevant stakeholders can solve problems associated with urban agglomerations (Baron, 2012). Therefore, SSCs valorize the capacity of clever people to generate solutions to solve urban issues. The people dimension is a crucial axis for the development of a city (Florida, 2002; Giffinger et al., 2007).

Welfare systems imply the correct use of technological systems. Internet and broadband network technologies as enablers of e-services become progressively more important for urban development as cities increasingly assume a critical role as drivers of innovation in fields such as health, inclusion, environment, and business (Kroes, 2010). The stimulation of ICT-based applications enhancing the quality of life for citizens is now becoming a key priority. Policymakers, citizens, and enterprises are primarily interested in concrete and short-term solutions that benefit business creation and social participation (Petrolo et al., 2015). While many cities have initiated ICT innovation programs to stimulate business and societal applications, scaling up pilot projects to large-scale, real-life deployment is nowadays crucial (Neirotti et al., 2014; Washburn et al., 2009).

Porter proposed a model aimed to introduce more efficient welfare systems based on four determinants: physical and immaterial infrastructure, networks and collaboration, entrepreneurial climate and business networks, demand for services, and availability of advanced end-users (Porter, 1990). Infrastructures for education and innovation, networks between businesses and governments, and intense demand for innovation and the quality of services are other relevant determinants of urban welfare. The national competitive advantage depends on the welfare system of urban areas (Porter, 1990).

The smart culture embraces urban cultural heritage, urban creative industries, citizen-centric vision, and the promotion of livability within cities (Allam & Newman, 2018; Piccialli & Chianese, 2018; Rutten, 2006). Culture is a driver for sustainable development. It is a pillar of urban regeneration and plays a relevant role in revamping urban areas (Rutten, 2006). In addition, culture affects the economic performance of urban units, promoting business development and job creation (Scott, 2004).

The preservation of the natural environment is a prominent issue in SSC (Bibri & Krogstie, 2017). An approach oriented to the optimal management of the natural urban resource depends on the policies adopted in different fields, such as energy, waste, and water landscape (Barrionuevo et al., 2012). An important strategy to favor the optimal management of natural resources should be associated with participatory policies. In this way, it is possible to leverage the citizens' awareness about the correct conservation of natural heritage and nonrenewable resources (Thuzar, 2011). These initiatives help create desirable conditions for a livable and sustainable city to preserve the natural environment and increase the city's attractiveness and livability (Alawadhi et al., 2012). The natural environment is a strategic component to sustain the future success of a smart city (Albino et al., 2015).

The ICT domain of SSC refers to the application of a wide range of electronic and digital technologies to create a cyber, digital, intelligent, wired, informational, or knowledge-based city. ICT-enabled applications and functionalities consist of the hardware, software, stable connectivity, interoperability, technology networks, availability of dashboards, common operational platform, integrated web services, control systems as automatic control networks, local operating networks, and big data (Bibri, 2018; Caragliu et al., 2009). In this regard, big data can transform every sector of a national economy (Batty, 2013). The potential deriving from big data is applicable in many ways to improve the SSC system; for example, it is possible to use sensors and wearable devices to monitor health levels more rapidly. Big data analysis benefits transport systems, generating efficiency, reducing congestion, and streamlining traffic automation processes. By exploiting the potential offered by smartphones, it is possible to enable smart governance aimed at problem-solving and greater transparency (Meijer & Bolívar, 2015; Willke, 2007). The use of information technology can transform mobility in a way to have broader effects on life and work. In other words, ICT can change the city infrastructure (Komninos, 2002; Komninos et al., 2011). Cities implementing emerging technologies are more likely to survive than laggards (Audretsch et al., 2021; Glaeser, 2005).

Knowledge workers esteem global cities (Sassen, 1991). Global cities show higher innovation rates, measured as patents and scientific production than non-global cities (Bettencourt et al., 2007; Schlapfer et al., 2014). Specifically, half of all inventions are born in global cities (Belderbos et al., 2020). In other words, most creative minds lie in global cities. It means that global cities tend to attract creative people. Global cities occupy a central position in the global scenario (Verginer & Riccaboni, 2021). Thanks to their location in the international mobility network and the higher level of infrastructure, amenities, and opportunities they can ensure, global cities are the ideal destination for scientists and inventors (Azoulay et al., 2017; Dong et al., 2020; Zacchia, 2018). Global cities attract early-stage career scientists and researchers that rationally select their future location. Two mechanisms affect the mobility of the scientists toward global cities: (1) spatial proximity tends to increase the citations; and (2) worldwide centrality of global cities sharpens the social proximity effect in a way to raise the visibility of the scientific production also at a long distance (Verginer & Riccaboni, 2021). Global cities also possess an established reputation, better research institutions and infrastructures, international enterprises, and interconnected local networks.

All these elements increase the SSC's appeal in the eyes of the people. Knowledge workers, creative people, and entrepreneurs, to name a few, search for friendly environments to live in, characterized by opportunities for a healthy life and successful career. When the city is based on a knowledge-based economy, the major attractiveness of a city captures more knowledge workers (Ivaldi et al., 2020). In this case, the feeling of working opportunities adds to the liveability (Qian & Acs, 2013). Knowledge workers can leverage their skills and creativity (Audretsch & Belitski, 2013) to find relevant opportunities in these contexts. Therefore, improving a city's strategies, policies, and infrastructures to increase its livability attracts more knowledge workers when the city's dominant economic model is knowledge-based.

2.4 From Knowledge to New Businesses Formation through Knowledge Workers

Knowledge and entrepreneurship are reciprocally dependent (Penco et al., 2020; Qian & Acs, 2013; Tavassoli et al., 2021). Leveraging the knowledge spillover theory of entrepreneurship (Acs et al., 2009; Audretsch & Belitski, 2013), they point to new knowledge as one source of entrepreneurial opportunities. People willing to assume entrepreneurial risks are potential entrepreneurs. They leverage their individual entrepreneurial orientation (Wales, 2016) to engage in innovative, proactive, and risky activities to undertake entrepreneurial activities through new firms. The latent potential of knowledge is made effective through the channel of entrepreneurship. The distinctive factor that allows knowledge to be introduced into the economic circuit, and to capitalize on it, is commercialization (Qian & Acs, 2013). Knowledge creation is not sufficient to achieve economic growth. Economic growth depends on the capability of the economic actors to capitalize on the new knowledge generated. If not commercialized (Arrow, 1962), the new knowledge does not generate profits (Teece, 2018). The commercialization of new products and services determines knowledge spillovers. New knowledge, in fact, does not automatically mean producing positive influences. It is necessary that knowledge dissemination occurs, that other creators can understand its potential, and that clients attribute a degree of usefulness to the commercialized product to initiate a spillover. Entrepreneurship is the conduit by which knowledge is exploited, disseminated, and capitalized. The literature identifies the actors involved in the commercialization of new knowledge as knowledge workers, attracted by cities having a high knowledge-based potential to be exploited (Qian & Acs, 2013; Romer, 1990). These workers must have a high propensity to undertake entrepreneurial activities. This attitude lies in individual-level characteristics. For instance, individual entrepreneurial orientation and self-efficacy represent two well-known constructs in business studies. They determine the ability of an entrepreneur to perform tasks. Individual entrepreneurial orientation indicates a personal risk-taking, innovative, and proactive attitude (Edmond & Wiklund, 2010; Miller, 2011). Individual entrepreneurial orientation is a strategic posture that influences the decisions about the future of the enterprise (Wales, 2016). Self-efficacy is the personal belief in her individual ability to accomplish a job or a specific set of tasks (Bandura, 1977). Self-efficiency is associated with the accumulation of consciousness during life. Problem-solving, relational capabilities, leadership, and decision-making depend on self-efficacy (Wilson et al., 2007). The KSAOs framework, i.e., knowledge, skills, abilities, and other characteristics, summarizes the fundamental attributes an entrepreneur should possess to perform her job (Ployhart & Moliterno, 2011). In general, entrepreneurial ability depends on the pool of applicable information to perform specific tasks and consists of foreign language knowledge or computer programming languages. Skills are a proxy of the degree of accuracy in performing a task. Abilities refer to more stable traits such as cognitive, sensory, and physical abilities (e.g., empathy). Other characteristics indicate a bundle of residual categories that involve certifications, degrees, personality, and values (Ployhart & Moliterno, 2011).

In summary, the knowledge spillover theory of entrepreneurship is based on two primary cornerstones: first, the not commercialized knowledge as a source of entrepreneurial opportunities; second, entrepreneurship, consisting in the act new firm's creation to extrapolate value from the knowledge (Qian & Acs, 2013). A direct relationship between knowledge and entrepreneurship clearly emerges as well as the sensibility concerning the level of entrepreneurship of an area to the marginal increment of the knowledge stock (Acs et al., 2009, p. 17).

Higher rates of new high-tech business entry and innovative performance characterize creative cities (Florida, 2011). The higher the new knowledge concentration of a city or region, the more intense the generation of entrepreneurial opportunities (Florida, 2004). Conversely, where new knowledge runs low, the consequences will be negative in terms of new entrepreneurial ideas and opportunities. This perspective shows a linkage between knowledge and entrepreneurship that inextricably go hand in hand.

Knowledge spillover leads to entrepreneurship creation under the condition of local institutional voids (Acs et al., 2013; Gallouj, 2017). The example of Detroit City teaches (Bendickson et al., 2021). After a flourishing period in the first half of the twentieth century, Detroit experienced economic setbacks that transformed the urban environment into a chaotic one. However, entrepreneurs adopted a prospect behavior of looking at the opportunities rather than potential threats and started to do business. Entrepreneurs developed their success stories around Detroit City's image. Supported by a sharpened media coverage, knowledge flowed among entrepreneurs. Informal institutions filled the formal institutional void (Bendickson et al., 2021; Bruton et al., 2010; Puffer et al., 2010). In other words, knowledge spillovers were means to overcome institutional voids. Entrepreneurs behaved as catalysts of such great potential through their interconnected networks. In Detroit City, knowledge and entrepreneurship represented the way a city survives and flourishes under the condition of the institutional void.

2.4.1 The Decisive Role of the Entrepreneurial Absorptive Capacity

The entrepreneurial absorptive capacity is a critical factor affecting the transmission of knowledge spillover of the entrepreneurs (Qian & Acs, 2013). They define the entrepreneurial absorptive capacity as "the ability of an entrepreneur to understand new knowledge, recognize its value, and subsequently commercialize it by creating a firm" (Qian & Acs, 2013, p. 191). While individual entrepreneurial orientation refers to the intention or propensity of a person to undertake an entrepreneurial activity, the entrepreneurial absorptive capacity is her capability to absorb and capitalize on the knowledge in a later step. In fact, it gets involved after the manifestation of will by an individual. It indicates a work-related capability, not a behavior. The entrepreneurial absorptive capacity explains the relationships between created knowledge, knowledge embodied in people (i.e., human capital), and entrepreneurship (Qian & Acs, 2013). Through the concept of entrepreneurial absorptive capacity, the authors indicate that individual-level factors (e.g., entrepreneurial absorptive capacity) affect knowledge spillover in a specific area (Qian & Acs, 2013). In other words, it tells us that new knowledge does not necessarily lead to entrepreneurship (Michelacci, 2002) if specific individual capabilities do not assist this process. More precisely, the discovery and exploitation of new knowledge are contingent on the entrepreneurial ability to sense opportunities and commercialize innovations into the market through ad hoc resource mobilizations.

2.4.2 Creativity as a Means to Pursue New Business Formation and Knowledge Filters Obstacles

Creativity refers to a process of introduction and implementation of new ideas that, because of their relevance, make other ideas useless and trivial (Florida, 2002). Creativity lies in an individual who adds economic value to a firm (Florida, 2002). The context is where the relationships occur and the interlocutor affects the individual propensity to produce creativity spillovers. Creativity (e.g., human capital and information) can only be transmitted when group members are more prone to jointly take risks, resolve uncertainty, and find new ways of problem-solving. The working environment of the creative classes should be non-threatening, tolerant, and open to different ideas and cultures (West, 1990). An environment full of new and alternative ideas, solutions, networks, and cultural diversity defines a melting pot environment (Bassett-Jones, 2005; Boschma & Fritsch, 2009; Lee et al., 2004). This environment enhances economic development in cities depending on the level of creativity. Diversity and cosmopolitanism are elements that can decisively raise the level of creativity (Audretsch et al., 2010).

Creativity depends on the ability to identify complex problems, formulate propositions, make inferences from hypotheses, and discuss ideas with others through socialization and networking. Consequently, when creative people engage in a creative process and transfer creative ideas, these ideas become marketable products. Problem identification and new idea sharing benefit creative activities, not just product creation (Drazin et al., 1999; Gilson & Shalley, 2004; Kahn, 1990).

Creativity and ideas are crucial components in enhancing the performance of organizations (Drazin et al., 1999; Fleming & Marx, 2006; Shalley et al., 2000, 2004; Somech & Drach-Zahavy, 2013) and regions (Acs & Armington, 2006; Rodrìguez-Pose & Vilalta-Bufì, 2005). Many studies have found a relationship between creativity and the level of entrepreneurship that characterizes a given area (Choi et al., 2009; Cohen & Bailey, 1997; Fleming et al., 2007; Gilson & Shalley, 2004; Shalley et al., 2004; Somech & Drach-Zahavy, 2013) or regions and countries

(Boschma & Fritsch, 2009; Fleming & Marx, 2006; Lorenzen & Andersen, 2009). A second relevant stream of inquiry refers to knowledge spillovers. They are greater in densely populated regions with a high degree of industrial density and cultural diversity (Audretsch et al., 2010), indicating exploitation of the competitive advantage of urban areas (Agarwal et al., 2007, 2010; Audretsch & Lehmann, 2005).

Creativity can be analyzed from different perspectives:

- Creativity, understood as a different way of measuring human capital, is heterogeneous and affects the level of entrepreneurship in the area (Choi et al., 2009; Fleming & Marx, 2006).
- The interaction between creativity and entrepreneurship accelerates urban economic development (Belitski & Desai, 2015a, 2015b). In other words, it is necessary to improve entrepreneurship to maximize the potential embedded in creativity. Enhancing entrepreneurship allows economic actors to carry creativity to generate urban economic development.
- Creativity is sensitive to cultural exchanges and commercialization. Cities characterized by high levels of openness, talent, and diversity are associated with higher development. The gap between commercialized and uncommercialized creativity is known as the creativity filter (Acs et al., 2009; Agarwal et al., 2010).

Entrepreneurship facilitates the spillover of creativity to urban economic development (Asheim & Hansen, 2009; Boschma & Fritsch, 2009; Florida, 2002). Given the close interdependence between context perception and creativity, it is justified to expect an increase in creative potential if there is a good perception of the environment in which entrepreneurial actors operate. Creativity is not a factor that affects urban economic development in the short term. However, it requires a channel for commercializing creative ideas (Audretsch & Belitski, 2013).

We must distinguish between ordinary and intellectual human capital. The former is a source of widespread and low-cost knowledge that ensures normal return rates (Acs et al., 2009; Zucker et al., 1998). The latter indicates the creative power embedded in the human capital (Boschma & Fritsch, 2009; Florida, 2002, 2004, 2011). Creativity lies in the intellectual human capital of the creative classes and generates return rates higher than normal (Florida, 2004). It can create a positive spiral of attraction, leading creative individuals to move to the most creative places. The literature calls this phenomenon self-selection of creative people in entrepreneurship (Acs & Megyesi, 2009). Areas of creativity proliferation are associated with the large-size cities that most show diversity and openness in culture, advanced technological architecture, and cultural amenities (Acs & Megyesi, 2009). These characteristics make a city more comfortable and contribute to attracting a large amount of educated, creative, and skilled workers. This pool of creative resources triggers regional growth by working with incumbents or starting new businesses. Revealing and valorizing a region's intellectual human capital is critical to improving the process of knowledge diffusion through transmitting personalized (tacit) knowledge.

2.5 The Triangulation between SSCs' Principles, Technological Advancement, and Knowledge-Based Economy

The formation of SSCs follows three steps: First, intelligent cities emerge as knowledge-based cities where institutional leadership and organizational capacity interplay to support technological advancement, innovation, and personal creativity (Angelidou, 2015). Intelligent cities aim to increase competitiveness and sustainability. Second, intelligent cities become smart cities emphasizing social and institutional architecture to assist policies and processes that provide an integrated and user-friendly urban configuration (Ojo et al., 2016). In smart cities, all the components coexist as an organic body (Albino et al., 2015; Nam & Pardo, 2011). Finally, the transition from smart cities to SSCs occurs along the means-end line. In SSCs, ICT is the means to achieve the purpose of sustainability (Kourtit & Nijkamp, 2012; Rivera et al., 2015). In other words, four forces shape SSCs, i.e., urban future, knowledge- and innovation-based economy, technology-push, and demand-pull solutions (Angelidou, 2015). Knowledge-based economy shapes the identity of smart cities toward their transformation into sustainable ones. The knowledge economy scaling implies new knowledge creation to find ad hoc solutions to the community's problems. The term sustainability, in fact, refers to meeting the current and future generations' needs without compromising environmental resources (Höjer & Wangel, 2015). Sustainability involves inter-generational issues also. Therefore, SSCs embrace both human and environmental challenges. SSC leverages new knowledge production to pursue multilevel objectives. For instance, SSC deploys technology to facilitate knowledge and innovation (Angelidou, 2015). These technology-pushed or demand-pulled processes of new knowledge creation depend on the interaction between multiple elements such as entrepreneurial, human, infrastructural, and social capital. Therefore, a direct link exists between a city's entrepreneurial system and knowledge creation capacity (Zheng et al., 2020). A city must struggle to attract knowledge and creative workers, form networks, and develop business clusters to improve its knowledge stock. We can interpret the relationship between a SSC and a knowledge-based economy as bijective. From one side, a SSC can influence the tendency of its knowledge-based economy to prosper by offering incentives to attract crucial actors (e.g., entrepreneurs). On the other side, these actors contribute to providing solutions to address needs, increasing the knowledge stock, and strengthening networks to generate positive feedback loops, i.e., enabling self-reinforcing processes of the city, entrepreneurial system, and knowledge-based economy (Fig. 2.1). In other words, knowledge is the engine of SSC since it determines the ability to undertake a people-first strategy (UN, 2015; Yigitcanlar et al., 2018, 2019).

Knowledge provides the key to the competitive advantage of firms, organizations, and cities (Angelidou, 2015). To increase the scale of its knowledge economy, SSC must deal with "glocal" tensions. In fact, the processes of knowledge creation and application are affected by glocal tensions. In fact, new knowledge production

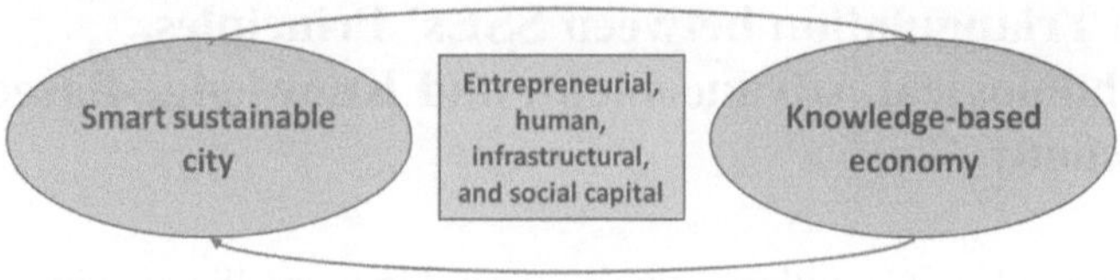

Fig. 2.1 Feedback loops between SSC and knowledge-based economy

occurs on a global scale while knowledge application takes place on a local scale. This factor explains why economic geography still plays a relevant role in the strategic design of the economy. Knowledge centrality is also demonstrated by the knowledge frameworks developed by international institutions, such as the European Union, the Organization for Economic Cooperation and Development, the United Nations, and the World Bank, to manage knowledge proliferation at global and local levels (Angelidou, 2015). Glocal tensions are due to recent technologies that have changed the production and management of knowledge in cities. Digital platforms have enabled long-distance collaboration, knowledge exchange networks, data storage, and analytics. Technologies (e.g., social networks) have enabled individuals to develop ideas and strategies and exchange opinions more easily (Komninos, 2002). A SSC is a complex body whose balance depends on the management of centrifugal thrusts seeking knowledge to be acquired and centripetal thrusts resulting from the localization of acquired knowledge.

2.6 The Relationships between Knowledge, Entrepreneurship, and Public Institutions

A knowledge-based SSC must leverage the interrelationships between knowledge production, entrepreneurial actors, creative people, knowledge workers, and public institutions to prosper. Specifically, the relationships between entrepreneurship and public institutions directly affect knowledge creation. These relationships can be either positive or negative, depending on whether they generate beneficial or detrimental effects in terms of knowledge production.

Creating a large base of intellectual human capital is a powerful driver of knowledge production in the entrepreneurial setting. However, individuals do not operate in a vacuum and sometimes can desist from starting a business. Despite their implicit entrepreneurial attitude, they seem to lack adequate incentives. From the perspective of the institutional setting, the concept of knowledge filter explains better than others why creative individuals often might decide not to start a business, even if they have adequate skills. The knowledge filter notion was formalized in the literature to indicate bureaucratic constraints, entrepreneurial opportunities, culture, regulatory barriers to entrepreneurship, social acceptance, and taxes that represent the most influential enablers or obstacles in terms of knowledge production and transmission through entrepreneurial activities (Acs et al., 2009; Parker, 2004; Stenholm et al., 2013).

A knowledge filter can have either a positive or negative meaning based on the type of filter and the role it plays in stimulating or reducing entrepreneurial dynamism. Undoubtedly, a positive effect is associated with the concept of organizational sponsorship. It is a mechanism of institutional arrangement (e.g., incubators) to assist new firm ventures (Amezcua et al., 2020). More specifically, organizational sponsorship serves as an institutional architecture to grow the economy, stimulate entrepreneurial activities, and improve the start-ups' likelihood of survival through founding and increasing new ventures' quality (Brixy et al., 2013). Amezcua et al. (2020) demonstrated that organizational sponsorship delays new venture exit under the condition of simultaneous low or high urbanization and localization. Urbanization indicates the concentration of economic activity in a specific area. Localization refers to the agglomeration of firms belonging to the same industry in a specific area (Moomaw, 1988). The economic activity concentration reduces the distance between firms and individuals and facilitates the searching for resources, their acquisition, and delivery. In summary, organizational sponsorship indicates positive interactions between institutions and entrepreneurship.

Public-private interactions can result in partnerships. Public-private partnerships refer to a structured form of collaboration between the public and private sectors. Public-private partnerships are enacted to finance, build, and operate large-scale projects (Public-Private Partnership Knowledge Lab, 2022a). For this reason, they often develop in the long term. These partnerships combine knowledge from the private sector and public sector incentives and regulatory power to pursue common objectives. The advantages of public-private partnerships are operational efficiency improvement or service provision, economic diversification, and risk reduction (The World Bank, 2022). As every activity involves forms of entrepreneurship, these partnerships are not without any risks or potential disadvantages. However, they represent the most viable solution to achieve specific objectives that benefit society and the entire city system. Particularly important for SSC is that public-private partnerships work for projecting and constructing environmental infrastructures and public service accommodations (Public-Private Partnership Knowledge Lab, 2022b). Therefore, they can be employed to improve the sustainability and urban quality of life levels.

2.7 Conceptual Framework

From the process perspective, the mechanism of knowledge spillover often departs with the generation of new knowledge by firms, research institutions, universities, and inventors as a result of creative activities (Fig. 2.2). Entrepreneurship is the conduit to provide market access to previously not commercialized knowledge. The linearity of this mechanism depends on individual- or organizational-level factors (e.g., entrepreneurial absorptive capacity). For instance, entrepreneurial absorptive

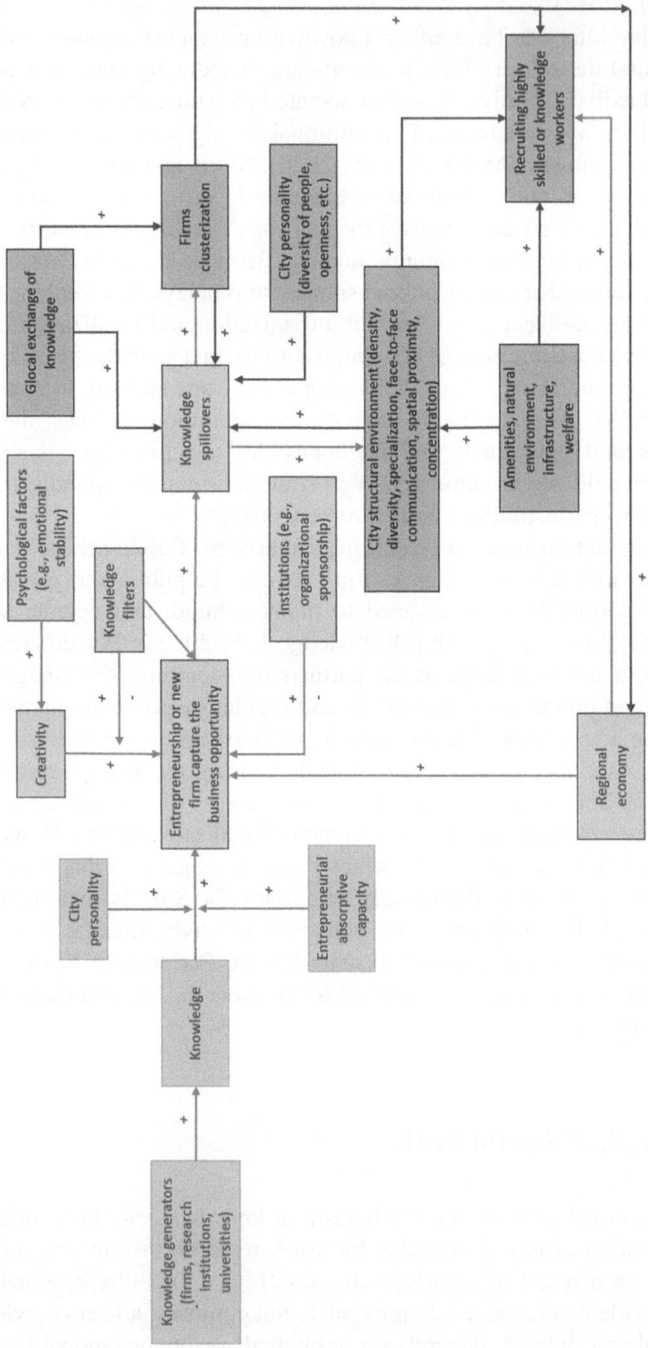

Fig. 2.2 Process of knowledge spillover formation and exploitation within SSC

capacity increases the individual or organizational ability to process information or apply knowledge in the business field. It leads individuals to create new firms to capitalize on the value of the knowledge absorbed. When new knowledge pairs with high levels of entrepreneurial absorptive capacity, the individual propensity to undertake business activities increases. It benefits the local entrepreneurial system and expands its potential. Entrepreneurship, i.e., new firms capturing business opportunities, is influenced by different factors at multiple levels. For instance, creativity is an element that benefits entrepreneurship since it identifies an individual attitude toward introducing and implementing new ideas. Despite creativity being a strong driver, it is not the only factor affecting entrepreneurship. The size of the regional economy and its growth trend affect entrepreneurship. The higher the size of the regional economy and its growth, the higher the propensity of the individuals to undertake entrepreneurial activities. The relationships with the public institutions, policies, specific actions to support local entrepreneurship (e.g., organizational sponsorship), and urban development strategies can stimulate or inhibit entrepreneurship. Knowledge filters accelerate or reduce the pace of entrepreneurship depending on their positive or negative nature. All these factors jointly determine the vitality of an entrepreneurial system of a city. Commercialized knowledge generates spillovers through the mediating actions of knowledge workers and entrepreneurs. New business formation directly allows the spreading of knowledge within a city's economic and industrial system and the commercialization of new knowledge. The literature has demonstrated that the relationship between knowledge spillover and firm clusterization is bijective. Knowledge spillover (or new knowledge creation) can lead to new business cluster formation. At the same time, clusters of firms strengthen the diffusion of knowledge. As we have previously written, the relationship between knowledge spillovers and clusters of firms is to be imagined as a mutual self-reinforcing process. In other words, considerable volumes of knowledge spillovers attract new firms that nurture the flow of knowledge spillovers by providing new knowledge, commercializing the existing one, or collaborating with partners to introduce new knowledge. The urban system must manage glocal tensions to capitalize on knowledge spillovers. These tensions require a balance between the need to oversee the global knowledge market and satisfy the production needs by establishing production facilities at the local level. Knowledge spillovers benefit firms' clusterization and improve the city's structural environment in terms of density, diversity, spatial proximity, and specialization. The combination of these industrial characteristics generates opportunities for reviving knowledge spillovers. In addition, a valuable and well-combined city structural environment attracts workers with various backgrounds. Attracting and recruiting knowledge workers is crucial to improve the likelihood of high-intensity knowledge spillovers. This trait is further enhanced when a robust industrial system and livable city (e.g., cultural amenities, healthy natural environment, welfare system) complement each other. Skilled knowledge workers are more likely to locate in more harmonious industrial and living systems. A SSC's value proposition is based on naturally and socially healthy environments, i.e., a symbiotic environment. Knowledge-based economy and SSC positively interact to keep going this balancing. The interdependence

Table 2.1 Factors influencing knowledge creation in an urban environment

Level	Types	Source
Individual	Creativity, individual entrepreneurial orientation, self-efficacy, absorptive capacity	Florida (2002), Boschma and Fritsch (2009), Wales (2016), Bandura (1977), Amezcua et al. (2020), Qian and Acs et al. (2013), Audretsch and Belitski (2013)
Organizational	Creativity, innovativeness, R&D, ICT, absorptive capacity, face-to-face communication, culture, strategic posture	Acs et al. (2009), Romer (1990), Wales (2016), Bibri (2018), Qian and Acs et al. (2013)
Institutional	Knowledge filters (bureaucracy, regulation, policy), strategies, organizational sponsorship (e.g., incubators), public-private partnership	Zheng et al. (2020); Angelidou (2015), Amezcua et al. (2020), Acs et al. (2009)
City	Structural environment (e.g., combination of density, diversity, specialization), personality, cluster, regional system, amenities, cultural diversity, technological architecture, ICT, mobility system, governance, face-to-face communication (e.g., coworking), tolerance	Tavassoli et al. (2021), Bibri and Krogstie (2017), Bibri (2018), Autio et al. (2014), Porter (1990, 1996), Delgado et al. (2014); Jacobs (1961, 1969); Audretsch et al. (2021)
Economic	Entrepreneurship, labor productivity, SMEs, high-tech, innovative firms, regional growth, taxation	Qian and Acs et al. (2013); Tavassoli et al. (2021); Acs et al. (2009); Audretsch and Belitski (2013)

between the knowledge economy and a SSC is functional to apply the concept of a "laboratory for innovation" city into facts (Batty et al., 2012, pp. 481–482).

From the previously presented theorizations, we can deduce that multilevel factors influence the creation of new knowledge (see Table 2.1). These factors simultaneously and synergistically determine knowledge production and its spillover through entrepreneurial actions (e.g., commercialization). We distinguish five-level factors that influence knowledge creation within urban environments. These levels refer to individuals, organizations, institutions, cities, and the economy. At the individual level, the attributes (e.g., creativity, entrepreneurial orientation, self-efficacy, absorptive capacity) associated with professionals, workers, and entrepreneurs directly influence their ability to create knowledge. They also enhance the propensity of the creative classes to start businesses and enter the market to commercialize new knowledge. Some individual attributes can be transferred to organizations (e.g., creativity, absorptive capacity, innovativeness) that produce new knowledge through team-working mechanisms. In addition, an organization's R&D, technological architecture, culture, and strategic posture determine the volume of new knowledge generated. Business organizations are nuclear components of the knowledge-based economy. The institutional level can have positive or negative effects on knowledge creation. For instance, knowledge filters can benefit the knowledge economy by incentivizing favorable policies for businesses.

However, an excess of bureaucracy and constraining regulations can discourage entrepreneurial activities, limit the creative freedom of individuals, and undermine the process of new knowledge generation. The institutions must adopt good strategies for improving the knowledge creation cycles within the urban area. They can also stimulate the proliferation of new businesses through organizational sponsorship and collaborate on large-scale programs with the private sector. The city's structural environment, personality, amenities, culture, and welfare system should support the location of knowledge workers and entrepreneurial activities. In this regard, clusters and regional innovation systems are relevant factors in attracting knowledge generators from the outside. Finally, the economy of a place stimulates the location of people and organizations that can benefit local knowledge creation while enjoying high standards of living.

2.8 SSCs' Strategies to Improve the Knowledge-Based Economy

Europe and North America show two different patterns of SSC development (Mora et al., 2017). The former represents a model built around a demand-pulled environmentally sustainable conception. The latter promotes technology advancement supported by high-tech firms such as IBM and Cisco (Mora et al., 2018). The European model lies in the cities' need to manage the complexity of urban living, such as energy and resource consumption, environmental protection, and overcrowding (European Parliament, 2014). The North American model demonstrates that the SSC future is strategically associated with the internal knowledge organization of a small number of global technology firms (McNeill, 2015). These differences show that SSCs can achieve similar results by adopting different strategies for urban development.

Angelidou (2015) indicates the strategies to achieve the integrated model of the smart city. These strategies consist in advancing human and social capital, adopting behavioral changes in terms of a sense of agency and meaning, and improving technological responsiveness to address the needs of the local community. Given the specific characteristics of SSCs, aligned to those of smart cities, these strategies can be borrowed and applied to the SSC context. However, in doing this, we must consider the distinctive trait of "sustainability" that characterizes the SSCs. City-level strategies must involve both technology and knowledge as pillars since technology supports knowledge creation and vice versa (Angelidou, 2015). The outcomes deriving from this interplay can benefit the natural environment. Bibri (2018) illustrates how ICT associated with sensors, big data analytics, platforms, and computing models finds applications in different domains of environmental sustainability in smart cities. For example, IoT helps cities monitor car emissions, air quality, and water quality. It can also reduce noise pollution. IoT enables machine learning and data mining to monitor atmospheric conditions, achieve higher environmental protection, and secure complex and interconnected urban environments against natural disasters (Bibri, 2018).

One of the most commonly suggested strategies refers to people-first strategies, i.e., strategies having urban livability and meeting human needs as pillars. People represent not only a driver to create new knowledge and improve the knowledge economy but also a target to assume in order to formulate clear and realistic objectives to be pursued through a city development strategy. Quality of life, intergenerational sustainability, and environmental sustainability are three pillars of SSCs. Their increasing leads to improve the city's reputation at a global level and attract more knowledge workers and creative individuals.

Policymakers should enlarge the boundaries of their governance strategy. Moving from an intra-city to a cross-city governance strategy allows policymakers to strengthen the inbound effect of attraction. Therefore, people-first strategies marry cross-city governance strategy as a continuum of aims. SSCs must manage their relationships with other cities as a lever of growth. They should adopt cross-city governance strategies to go beyond the boundaries of intra-city governance and acquire new knowledge, attract human and financial capital, and improve the entrepreneurial system (European Parliament, 2014; Zheng et al., 2020).

The entire cycle of knowledge management, from creation to exploitation, is affected by glocal tensions, i.e., the actions of multiple factors both at the global and local level that shape knowledge-related trajectories. SSCs must adopt inside-out strategies to manage glocal tensions. SSCs cannot avoid adopting technologies such as platforms to operate in the worldwide knowledge market. Digital platforms enable large-scale collaboration, information collection, and knowledge exchange (Komninos, 2002). At the same time, SSC must exploit knowledge at the local level. Regional clusters and regional systems of innovation generate knowledge spillover, knowledge refinement, and new knowledge creation (Komninos, 2002). They allow the formation of new firms with positive effects in terms of employment, economic growth, and local resources valorization (e.g., human capital). Therefore, operating both in the global and local knowledge market is not a mutually alternative choice but a competitive and strategic move to maximize the complementarity between different sources of knowledge. Leveraging complementary channels and sources of knowledge is a way to produce knowledge in combination with a high value-added.

Citizens' involvement, engagement, and participation imply that relationships with multiple stakeholders are established and managed. This approach is part of public-private partnership development. These partnerships involve knowledge and resource sharing by public and private actors to pursue common objectives. Public-private partnerships make research and innovation more responsive to local needs (OECD, 2016; Zheng et al., 2020). In addition, they ensure protection against potential risks during the developmental stages. In fact, resource sharing means that adverse effects are managed more flexibly by different actors. In other words, public-private partnerships are strategically relevant in risk management. In general, risk management is an iterative process involving five steps, i.e., identification, analysis, evaluation, treatment, and monitoring and control of risks (Ullah et al.,

2021; Wang et al., 2010). Risks arise from multiple sources. For instance, technology is a controversial element. In fact, technology is both the tool through which policymakers can monitor, manage, or prevent risks and a potential source of risks. Technology exposes the complex SSCs' system to risks deriving from smartness. Pervasive connectivity is the key to implementing more sophisticated technology-enabled functionalities at the city level. At the same time, technology represents the key to access to infrastructures for hackers (Zheng et al., 2020). Another controversial element is big data. Big data provides valuable information to improve the management of the urban system. However, a data breach allows attackers to obtain strategically relevant and significant information. Therefore, policymakers must increase their ability to conceive and manage these "double-edged weapons." Other potential sources of risks refer to the organizations and the external environment (Ullah et al., 2021). Strategies to manage risks represent opportunities to minimize downsides and cascade effects on interconnected components of a city. In addition, they strengthen the city governance, safety, security, and privacy of citizens (Ullah et al., 2021).

2.9 Practical Implications

Based on the previous arguments, this paragraph formulates suggestions for practitioners. In general, knowledge-based smart sustainable cities' public decision-makers should promote policies aimed at stimulating local knowledge spillovers and economic vitality. These policies should affect the city's structural environment, human capital, and social capital. In addition, they should consider the local psychological tendencies, i.e., the openness of people (Huggins & Thompson, 2019; Shane, 2003).

In addition, policymakers should concentrate on the city's personality traits. For instance, public policies could develop tools to enhance the openness of the local communities and promote intellectual and cultural growth (Zimmermann & Neyer, 2013), enable people to share new ideas, and tolerate diverse ideas (Tasselli et al., 2018). In other words, policymakers should launch awareness campaigns among city residents. Other policies should aim to deploy the potential of young people (e.g., young students) by stimulating their interaction and structuring a favorable city environment that brings them together (density) and offers new ideas and possibilities (diversity) (Tavassoli et al., 2021).

Entrepreneurs make informed choices about where to locate their business (Saxenian, 2007; Walshok & Shragge, 2013) and live. They can consider different factors, such as the most profitable areas, strong business networks or clusters, and the presence of good schools. Public institutions should promote actions to transform less entrepreneurial cities and regions to reduce disparities. Strategies to attract open and creative people can enhance the appeal of a specific area (Tavassoli et al., 2021).

Gentrification, studentification, and youthification determine the presence of young and highly educated people in the city. Gentrification indicates the process of change in a neighborhood due to new residents and businesses. Studentification consists in the influence that post-secondary students bring in a place. Youthification is the increasing concentration of young adults within a city. The simultaneous presence of these factors is critical for the knowledge-based economy success. Authorities should ensure a combination of housing and labor market facilitation to sustain these trends (Moos, 2016; Moos et al., 2019).

Many authors (Zait, 2017; Zheng et al., 2020) highlight that the human dimension of SSC has received limited attention compared to the technological section. It implies the number of studies on civil culture, individual competencies, soft skills, and behavioral characteristics should increase. At the same time, the choices of the policymakers must consider these aspects to formulate programs around the intellectual, human, and social capital of the city.

2.10 Conclusions

This chapter aims to clarify how a knowledge-based economy nestles in a SSC. The chapter focuses on the mutual relationship between the city and its knowledge economy to unpack some of the most relevant entrepreneurial, human, and social aspects. This chapter does not yearn to exhaustively cover complex issues such as the interaction between the urban environment and its economy. The chapter instead highlights that in such complex environments shaped by global pressures, one cannot avoid considering the interdependence between the city and economy. Envisioning the future of a SSC means considering the potential implications of the knowledge-based economy and vice versa. The chapter identifies the processes of knowledge spillover and the mutual influence of the elements of the city. In addition, it identifies the multilevel actors influencing knowledge creation in a SSC and the city's most influential factors of attraction. It also defines strategies through which policymakers can address SSC needs by leveraging knowledge and collaborating with various stakeholders. Overall, the interaction of different SSC components generates a circuit of knowledge transmission, i.e., an ecosystem of knowledge creation. This process benefits the knowledge-based economy and increases the quality of life that satisfies the living needs of the creative classes. In this regard, knowledge plays a central role in the value creation process of a SSC.

Montreal
Country: Canada
Inhabitants: 1,762,949
Density: 4828.3/km^2 (12,505/sq. mi)
Total area: 431.50 km^2 (166.60 sq. mi)

(continued)

The largest city in the Canadian province of Quebec, Montreal, is the Smart City case study in this book which is mostly focused on improving people's quality of life. The transformation affecting numerous industries (ICT, mobility, business, real estate, etc.) is inevitably changing the urban fabric of the city; it is during this period of profound evolution that local government plans to strategically intervene to lay the foundations for the creation of a future real Smart City.

The city of Montreal has decided to adopt, like other cities around the world, the so-called vision zero program, which expresses the desire to end traffic-related injuries and fatalities. To this end, a special committee composed of partners with extensive experience in road safety was established. The Canadian municipality also plans to issue an annual road safety report that includes data, KPIs, and mitigation strategies, focusing, in particular, on intersection crossings, speed limits, and the transit of large vehicles through urban areas.

The construction of additional bike lanes and the expansion of the current REV ("Réseau express vélo") network contribute to increased road safety. By the time this project is finished, 184 km of cycling routes should connect every part of the city along 17 different axes.

As abovementioned, improving people's Well-being is at the core of Montreal's sustainable strategy: The administration has chosen to pursue this goal by putting nature at the center of the project, both by establishing new green areas in the city and by safeguarding existing ones. In order to protect green spaces in urban areas and prevent overbuilding, the city has already adopted policies that have been in place for several years. In particular, the city has established some biodiversity centers, separated from one another and connected by protected natural corridors to allow migrating animal species to pass through. The city of Montreal has always been actively involved in climate change and has sought to establish itself as a world leader. Montreal has consistently met its pollution reduction targets through multiple initiatives to monitor and reduce the polluting emissions of the city's major operating sectors. The city of Montreal's commitment to the fight against climate change is also evident in its adherence to a number of international agreements, including the global covenant of mayors, the carbon disclosure project, and the C40. These agreements obligate the government to take specific and strict actions. The presence of community gardens where locals can grow their vegetables in a shared environment, bee protection rules that forbid the use of pesticides in outdoor urban places, and financing for scientific research on urban forests are all examples of local initiatives aimed at protecting nature.

From the entrepreneurial point of view, Montreal supports entrepreneurial activities through low costs of doing business, tax incentives, technologically advanced ecosystems, and conducive institutional environments. Also,

(continued)

Montreal City benefits innovative start-ups by offering capital financing, incubators, accelerators, and many investors. The evaluation formulated by the OECD, indicating Canada as the most attractive country for entrepreneurs worldwide, supports this perspective.

In addition, Montreal City offers a valuable environment for future prospects in the digital economy. This makes Montreal a city with a strong human and knowledge capital where a large pool of talent prospers. Entrepreneurs can recruit highly skilled professionals and knowledge workers can find many career opportunities by working remotely or in the multiple collaborative workspaces that Montreal displaces.

The high quality of life and knowledge-based economy make Montreal a perfect example of a cutting-edge city. In Montreal, the mutual self-reinforcing processes between knowledge spillovers and clusters of firms jointly contribute to creating value for the city.

References

Acs, Z. J., & Armington, C. (2006). *Entrepreneurship, agglomeration and US regional growth*. Cambridge University Press.

Acs, Z. J., Audretsch, D., & Lehmann, E. (2013). The knowledge spillover theory of entrepreneurship. *Small Business Economics, 39*(2), 289–300.

Acs, Z. J., Braunerhjelm, P., Audretsch, D. B., & Carlsson, B. (2009). The knowledge spillover theory of entrepreneurship. *Small Business Economics, 32*(1), 15–30.

Acs, Z. J., & Megyesi, M. I. (2009). Creativity and industrial cities: A case study of Baltimore. *Entrepreneurship and Regional Development, 21*, 421–439.

Acs, Z. J., & Sanders, M. (2012). Patents, knowledge spillovers and entrepreneurship. *Small Business Economics, 39*(4), 801–817.

Agarwal, R., Audretsch, D. B., & Sarkar, M. (2007). The process of creative construction: Knowledge spillovers, entrepreneurship and economic growth. *Strategic Entrepreneurship Journal, 1*, 263–286.

Agarwal, R., Audretsch, D. B., & Sarkar, M. (2010). Knowledge spillovers and strategic entrepreneurship. *Strategic Entrepreneurship Journal, 4*, 271–283.

Agrawal, A., Cockburn, I., Galasso, A., & Oettl, A. (2014). Why are some regions more innovative than others? The role of small firms in the presence of large labs. *Journal of Urban Economics, 81*, 149–165.

Ahvenniemi, H., Huovila, A., Pinto-Seppä, I., & Airaksinen, M. (2017). What are the differences between sustainable and smart cities? *Cities, 60*, 234–245.

Alawadhi, S., Aldama-Nalda, A., Chourabi, H., Gil-Garcia, J. R., Leung, S., Mellouli, S., Nam, T., Pardo, T. A., Scholl, H. J., & Walker, S. (2012). Building understanding of Smart City initiatives. In H. J. Scholl et al. (Eds.), *Electronic government* (pp. 40–53). EGOV 2012, LNCS 7443.

Albino, V., Berardi, U., & Dangelico, R. M. (2015). Smart cities: Definitions, dimensions, performance, and initiatives. *Journal of Urban Technology, 22*, 3–21.

Allam, Z., & Newman, P. (2018). Redefining the Smart City: Culture, metabolism and governance. *Smart Cities, 1*, 4–25.

Amezcua, A., Ratinho, T., Plummer, L. A., & Jayamohan, P. (2020). Organizational sponsorship and the economics of place: How regional urbanization and localization shape incubator outcomes. *Journal of Business Venturing, 35*(4), 105967.

Angelidou, M. (2015). Smart cities: A conjuncture of four forces. *Cities, 47*, 95–106.

Arrow, K. J. (1962). The economic implications of learning by doing. *Review of Economic Studies, 29*(3), 155–173.

Asheim, B., & Hansen, H. K. (2009). Knowledge bases, talents, and contexts: On the usefulness of the creative class approach in Sweden. *Economic Geography, 85*, 425–442.

Audretsch, D. B., & Belitski, M. (2013). The missing pillar: The creativity theory of knowledge spillover entrepreneurship. *Small Business Economics, 41*(4), 819–836.

Audretsch, D. B., Belitski, M., & Korosteleva, J. (2021). Cultural diversity and knowledge in explaining entrepreneurship in European cities. *Small Business Economics, 56*(2), 593–611.

Audretsch, D. B., Dohse, D., & Niebuhr, A. (2010). Cultural diversity and entrepreneurship: a regional analysis for Germany. *The Annals of Regional Science, 45*(1), 55–85.

Audretsch, D. B., Heger, D., & Veith, T. (2015). Infrastructure and entrepreneurship. *Small Business Economics, 44*(2), 219–230.

Audretsch, D. B., & Lehmann, E. E. (2005). Does the knowledge spillover theory of entrepreneurship hold for regions? *Research Policy, 34*(8), 1191–1202.

Autio, E., Kenney, M., Mustar, P., Siegel, D. S., & Wright, M. (2014). Entrepreneurial innovation: The importance of context. *Research Policy, 43*(7), 1097–1108.

Azoulay, P., Ganguli, I., & Graff Zivin, J. (2017). The mobility of elite life scientists: Professional and personal determinants. *Research Policy, 46*(3), 573–590.

Bandura, A. (1977). Self-efficacy: Toward a unifying theory of behavioral change. *Psychological Review, 84*(2), 191.

Baron, R. A. (2012). *Entrepreneurship: An evidence–based guide*. Edward Elgar.

Barrionuevo, J. M., Berrone, P., & Ricart, J. E. (2012). Smart cities, sustainable progress: Opportunities for urban development. *IESE Insight, 14*, 50–57.

Bassett-Jones, N. (2005). The paradox of diversity management, creativity and innovation. *Creativity and Innovation Management, 14*(2), 169–175.

Bateson, G. (1973). *Steps to an ecology of mind*. Paladin.

Batty, M. (2013). Big data, smart cities and city planning. *Dialogues in Human Geography, 3*(3), 274–279.

Batty, M., Axhausen, K. W., Giannotti, F., Pozdnoukhov, A., Bazzani, A., Wachowicz, M., Ouzounis, G., & Portugali, Y. (2012). Smart cities of the future. *The European Physical Journal, 214*, 481–518.

Belderbos, R., Benoit, F., Edet, S., Lee, G. H., & Riccaboni, M. (2020). Global cities' innovation network. In D. Castellani, A. Perri, V. Scalera, & A. Zanfei (Eds.), *Crossborder innovation in a changing world. Players, places and policies*. Oxford University Press.

Belitski, M., & Desai, S. (2015a). What drives ICT clustering in European cities? *Journal of Technology Transfer, 41*(3), 430–450.

Belitski, M., & Desai, S. (2015b). Creativity, entrepreneurship and economic development: City-level evidence on creativity spillover of entrepreneurship. *Journal of Technology Transfer, 41*, 1354–1376.

Bendickson, J. S., Irwin, J. G., Cowden, B. J., & McDowell, W. C. (2021). Entrepreneurial ecosystem knowledge spillover in the face of institutional voids: Groups, issues, and actions. *Knowledge Management Research and Practice, 19*(1), 117–126.

Bettencourt, L. M., Lobo, J., & Strumsky, D. (2007). Invention in the city: Increasing returns to patenting as a scaling function of metropolitan size. *Research Policy, 36*(1), 107–120.

Bibri, S. E. (2015). *The shaping of ambient intelligence and the internet of things: Historico-epistemic, socio-cultural, politico-institutional and eco-environmental dimensions*. Springer-Verlag.

Bibri, S. E. (2018). The IoT for smart sustainable cities of the future: An analytical framework for sensor-based big data applications for environmental sustainability. *Sustainable Cities and Society, 38*, 230–253.

Bibri, S. E., & Krogstie, J. (2017). Smart sustainable cities of the future: An extensive interdisciplinary literature review. *Sustainable Cities and Society, 31*, 183–212.

Boschma, R., & Fritsch, M. (2009). Creative class and regional growth: Empirical evidence from seven European countries. *Economic Geography, 85*, 391–423.

Brixy, U., Sternberg, R., & Stüber, H. (2013). Why some nascent entrepreneurs do not seek professional assistance. *Applied Economics Letters, 20*(2), 157.

Bruton, G. D., Ahlstrom, D., & Li, H. L. (2010). Institutional theory and entrepreneurship: Where are we now and where do we need to move in the future? *Entrepreneurship Theory and Practice, 34*(3), 421–440.

Caragliu, A., Bo, C. D., & Nijkamp, P. (2009). Smart cities in Europe. In *3rd Central European Conference in Regional Science* (pp. 45–60).

Choi, J. N., Anderson, T. A., & Veillette, A. (2009). Contextual inhibitors of employee creativity in organizations: The insulating role of creative ability. *Group and Organization Management, 34*, 330–357.

Cohen, S. G., & Bailey, D. E. (1997). What makes teams work: Group effectiveness research from the shop floor to the executive suite. *Journal of Management, 23*, 239–290.

Dameri, R. P. (2012). Searching for smart city definition: A comprehensive proposal. *International Journal of Computers and Technology, 11*, 2544–2551.

Delgado, M., Porter, M. E., & Stern, S. (2014). Clusters, convergence, and economic performance. *Research Policy, 43*(10), 1785–1799.

Diamond, J. (2005). *Collapse: How societies choose to fail or succeed*. Penguin Group USA.

Dirks, S., & Keeling, M. (2009). *A vision of smarter cities: How cities can Lead the way into a prosperous and sustainable future*. IBM Global Business Services.

Dixit, A. (2009). Governance institutions and economic activity. *The American Economic Review, 99*(1), 5–24.

Dong, X., Zheng, S., & Kahn, M. E. (2020). The role of transportation speed in facilitating high skilled teamwork across cities. *Journal of Urban Economics, 115*, 103212.

Drazin, R., Glynn, M. A., & Kazanjian, R. K. (1999). Multilevel theorizing about creativity in organizations: A sensemaking perspective. *Academy of Management Review, 24*, 286–307.

Edmond, V. P., & Wiklund, J. (2010). The historic roots of entrepreneurial orientation research. In H. Landstrom & F. Lohrke (Eds.), *Historic foundations of entrepreneurship research* (pp. 142–160). Edward Elgar Publishing.

European Parliament. (2014). *Mapping smart cities in the EU*. European Parliament.

Fleming, L., Chen, D., & Mingo, S. (2007). Collaborative brokerage, generative creativity, and creative success. *Administration Science Quarterly, 52*, 443–475.

Fleming, L., & Marx, M. (2006). Managing creativity in a small world. *California Management Review, 48*, 6–27.

Florida, R. (2002). *The rise of the creative class*. Basic Books.

Florida, R. L. (2004). *Cities and the creative class*. Routledge.

Florida, R. (2010). *Who's your city?: How the creative economy is making where to live the most important decision of your life*. Basic Books.

Florida, R. L. (2011). *A floating silicon valley for techies without green cards*. Accessed Sep 16, 2022, from https://www.bloomberg.com/news/articles/2011-12-02/a-floating-silicon-valley-for-techies-without-green-cards

Gallouj, F. (2017). Knowledge spillover-based strategic entrepreneurship. *International Journal of Entrepreneurial Behavior and Research, 23*(4), 726–730.

Gambardella, A., & Giarratana, M. S. (2010). Localized knowledge spillovers and skill-biased performance. *Strategic Entrepreneurship Journal, 4*(4), 323–339.

Giffinger, R., Fertner, C., Kramar, H., Kalasek, R., Pichler-Milanovic, N., & Meijers, E. (2007). *Smart cities: Ranking of European medium-sized cities*. Vienna University of Technology.

Gilson, L. L., & Shalley, C. E. (2004). A little creativity goes a long way: An examination of teams' engagement in creative processes. *Journal of Management, 30*(4), 453–470.
Glaeser, E. L. (2005). Reinventing Boston: 1630–2003. *Journal of Economic Geography, 5*(2), 119–153.
Glaeser, E. L. (2011). *Triumph of the city: How our greatest invention makes US richer, smarter, greener, healthier and happier*. Pan Macmillan.
Glaeser, E. L., Kolko, J., & Saiz, A. (2001). Consumer city. *Journal of Economic Geography, 1*(1), 27–50.
Glaeser, E. L., Saiz, A., Burtless, G., & Strange, W. C. (2004). The rise of the skilled city. *Brookings-Wharton Papers on Urban Affairs, 2004*, 47–105.
Globerman, S., & Shapiro, D. (2002). Global foreign direct investment flows: The role of governance infrastructure. *World Development, 30*(11), 1899–1919.
Höjer, M., & Wangel, J. (2015). Smart sustainable cities: Definition and challenges. In *ICT innovations for sustainability* (pp. 333–349). Springer.
Huggins, R., & Thompson, P. (2019). The behavioural foundations of urban and regional development: Culture, psychology and agency. *Journal of Economic Geography, 19*(1), 121–146.
International Telecommunications Union (ITU). (2014, March 5-6). *Agreed definition of a smart sustainable city, Focus Group on Smart Sustainable Cities*, SSC–0146 version.
Ivaldi, E., Penco, L., Isola, G., & Musso, E. (2020). Smart sustainable cities and the Urban knowledge-based economy: A NUTS3 level analysis. *Social Indicators Research, 150*, 45–72.
Jacobs, J. (1961). *The death and life of great American cities*. Vintage Books.
Jacobs, J. (1969). *The economy of cities*. Random House, Inc.
Jaffe, A. B. (1986). Technological opportunity and spillovers of R&D: Evidence from firms' patents, profits, and market value. *American Economic Review, 76*(5), 984–1001.
Johnson-Laird. (1983). *Mental models*. Cambridge University Press.
Kahn, W. A. (1990). Psychological conditions of personally engagement and disengagement at work. *Academy of Management Journal, 33*, 692–724.
Kanter, R. M., & Litow, S. S. (2009). *Informed and interconnected: A manifesto for smarter cities* (pp. 09–141). Harvard Business School General Management Unit Working Paper.
Klarl, T. (2013). Comment on Acs and Varga: Entrepreneurship. *Agglomeration and Technological Change, Small Business Economics, 41*, 215–218.
Komninos, N. (2002). *Intelligent cities: Innovation, knowledge systems and digital spaces*. Taylor & Francis.
Komninos, N., Schaffers, H., and Pallot, M. (2011). *Developing a policy road map for smart cities and the future internet*. E-challenges e-2011. Conference proceedings Paul Cunningham and Miriam Cunningham (Eds.). IIMC International Information Management Corporation.
Kourtit, K., & Nijkamp, P. (2012). Smart cities in the innovation age. *Innovation-Abingdon, 25*(2, SI), 93–95.
Kroes, N. (2010). *European commissioner for digital agenda. The critical role of cities in making the digital agenda a reality. Closing speech to global cities dialogue*. Spring Summit of Mayors Brussels.
Lee, S. Y., Florida, R. L., & Acs, Z. J. (2004). Creativity and entrepreneurship: A regional analysis of new firm formation. *Regional Studies, 38*, 879–891.
Lorenzen, M., & Andersen, K. V. (2009). Centrality and creativity: Does Richard Florida's creative class offer new insights into urban hierarchy? *Economic Geography, 85*, 363–390.
Marshall, A. (1920). *Principles of economics*. Macmillan.
McNeill, D. (2015). Global firms and smart technologies: IBM and the reduction of cities. *Transactions of the Institute of British Geographers, 40*(4), 562–574.
Meijer, A., & Bolívar, M. P. R. (2015). Governing the smart city: A review of the literature on smart urban governance. *International Review Administration Science, 82*, 392–408.
Michelacci, C. (2002). Low returns in R&D due to the lack of entrepreneurial skills. *The Economic Journal, 113*, 207–225.

Miller, D. (2011). Miller (1983) revisited: A reflection on EO research and some suggestions for the future. *Entrepreneurship: Theory & Practice, 35*(5), 873–894.
Moomaw, R. L. (1988). Agglomeration economies: Localization or urbanization? *Urban Studies, 25*(2), 150–161.
Moos, M. (2016). From gentrification to youthification? The increasing importance of young age in delineating high-density living. *Urban Studies, 53*(14), 2903–2920.
Moos, M., Revington, N., Wilkin, T., & Andrey, J. (2019). The knowledge economy city: Gentrification, studentification and youthification, and their connections to universities. *Urban Studies, 56*(6), 1075–1092.
Mora, L., Bolici, R., & Deakin, M. (2017). The first two decades of smart-city research: A bibliometric analysis. *Journal of Urban Technology, 24*(1), 3–27.
Mora, L., Deakin, M., & Reid, A. (2018). Combining co-citation clustering and text-based analysis to reveal the main development paths of smart cities. *Technological Forecasting and Social Change, 142*(SI), 56–69.
Nam, T., & Pardo, T. A. (2011). Conceptualizing smart city with dimensions of technology, people, and institutions. In *12th Annual International Digital Government Research Conference: Digital government innovation in challenging times* (pp. 282–291).
Neirotti, P., De Marco, A., Cagliano, A. C., Mangano, G., & Scorrano, F. (2014). Current trends in smart city initiatives: Some stylised facts. *Cities, 38*, 25–36.
Newman, P., Kosonen, L., & Kenworthy, J. (2016). Theory of urban fabrics: Planning the walking, transit/public transport and automobile/motor car cities for reduced car dependency. *Town Planning Review, 87*(4), 429–458.
Nonaka, I. (1994). A dynamic theory of organizational knowledge creation. *Organization Science, 5*(1), 14–37.
Obschonka, M., Stuetzer, M., Rentfrow, P. J., Shaw-Taylor, L., Satchell, M., Silbereisen, R. K., & Gosling, S. D. (2018). In the shadow of coal: How large-scale industries contributed to present-day regional differences in personality and Well-being. *Journal of Personality and Social Psychology, 115*(5), 903.
OECD. (2016). *Strategic public-private partnerships in science, technology and innovation*. Accessed Sep 16, 2022, from https://doi.org/10.1787/sti_in_outlook-2016-10-en
Ojo, A., Dzhusupova, Z., & Curry, E. (2016). Exploring the nature of the smart cities research landscape. In J. R. Gil-Garcia, T. A. Pardo, & T. Nam (Eds.), *Smarter as the new Urban agenda: a comprehensive view of the 21st Century City* (pp. 23–47). Springer International Publishing.
Parker, S. C. (2004). *The economics of self-employment and entrepreneurship*. Cambridge University Press.
Penco, L., Ivaldi, E., Bruzzi, C., & Musso, E. (2020). Knowledge-based urban environments and entrepreneurship: Inside EU cities. *Cities, 96*, 102443.
Pervin, L. A., & John, O. P. (1999). *Handbook of personality: Theory and research*. Elsevier.
Petrolo, R., Loscrì, V., & Mitton, N. (2015). Towards a smart city based on cloud of things, a survey on the smart city vision and paradigms. *Transactions on Emerging Telecommunications Technologies, 28*, 1–11.
Piccialli, F., & Chianese, A. (2018). *Editorial for FGCS special issue: The internet of cultural things: Towards a smart cultural heritage* (Vol. 81, p. 514). Elsevier.
Ployhart, R. E., & Moliterno, T. P. (2011). Emergence of the human capital resource: A multi-level model. *Academy of Management Review, 36*, 127–150.
Polanyi, M. (1966). *The tacit dimension*. Routledge & Kegan Paul.
Porter, M. E. (1990). *The competitive advantage of nations*. Free Press.
Porter, M. E. (1996). Competitive advantage, agglomeration economies, and regional policy. *International Regional Science Review, 19*(1–2), 85–90.
Porter, M. E. (2000). Location, competition, and economic development: Local clusters in a global economy. *Economic Development Quarterly, 14*(1), 15–34.

Public-Private Partnership Knowledge Lab. (2022a). *PPP contract types and terminology*. Accessed Sep 12, 2022, from https://ppp.worldbank.org/public-private-partnership/ppp-knowledge-lab

Public-Private Partnership Knowledge Lab. (2022b). *How PPPs Are Used: Sectors and Services*. Accessed Sep 12, 2022, from https://ppp.worldbank.org/public-private-partnership/ppp-knowledge-lab

Puffer, S. M., McCarthy, D. J., & Boisot, M. (2010). Entrepreneurship in Russia and China: The impact of formal institutional voids. *Entrepreneurship Theory and Practice, 34*(3), 441–467.

Qian, H., & Acs, Z. J. (2013). An absorptive capacity theory of knowledge spillover entrepreneurship. *Small Business Economics, 40*(2), 185–197.

Rentfrow, P. J., Gosling, S. D., & Potter, J. (2008). A theory of the emergence, persistence, and expression of geographic variation in psychological characteristics. *Perspectives on Psychological Science, 3*(5), 339–369.

Rentfrow, P. J., & Jokela, M. (2016). Geographical psychology: The spatial organization of psychological phenomena. *Current Directions in Psychological Science, 25*(6), 393–398.

Rivera, M. B., Eriksson, E., & Wangel, J. (2015). ICT practices in smart sustainable cities in the intersection of technological solutions and practices of everyday life. In *29th International Conference on Informatics for Environmental Protection/3rd International Conference on ICT for Sustainability* (pp. 317–324). ACSR-Advances in Computer Science Research.

Robertson, M., & Swan, J. (2003). Control–what control? Culture and ambiguity within a knowledge intensive firm. *Journal of Management Studies, 40*(4), 831–858.

Roche, M. P. (2019). Taking innovation to the streets: Microgeography, physical structure and innovation. *Review of Economics and Statistics*, 1–47.

Rodrìguez-Pose, A., & Vilalta-Bufì, M. (2005). Education, migration, and job satisfaction: The regional returns of human capital in the EU. *Journal of Economic Geography, 5*, 545–566.

Romer, P. M. (1986). Increasing returns and long-run growth. *Journal of Political Economy, 94*(5), 1002–1037.

Romer, P. M. (1990). Endogenous technological change. *The Journal of Political Economy, 98*, 71–102.

Rutten, P. (2006). Cultural activities & creative industries. A driving force for urban regeneration. In *Culture & Urban Regeneration; Finding & Conclusions on the economic perspective; Urbact culture network.*

Sassen, S. (1991). *The global city: New York, London, Tokyo*. Princeton University Press.

Saxenian, A. (2007). *The new argonauts: Regional advantage in a global economy*. Harvard University Press.

Schlapfer, M., Bettencourt, L. M. A., Grauwin, S., Raschke, M., Claxton, R., Smoreda, Z., West, G. B., & Ratti, C. (2014). The scaling of human interactions with city size. *Journal of the Royal Society Interface, 11*(98), 20130789.

Schumpeter, J. (1934). *The theory of economic development*. Oxford University Press.

Scott, A. J. (2004). Cultural-products industries and urban economic development: Prospects for growth and market contestation in global context. *Urban Affairs Review, 39*, 461–490.

Shalley, C. E., Gilson, L. L., & Blum, T. C. (2000). Matching creativity requirements and the work environment: Effects on satisfaction and intent to turnover. *Academy of Management Journal, 43*, 215–224.

Shalley, C. E., Zhou, J., & Oldham, G. R. (2004). The effects of personal and contextual characteristics on creativity: Where should we go from here? *Journal of Management, 30*(6), 933–958.

Shane, S. (2003). *A general theory of entrepreneurship: The individual-opportunity nexus*. Edward Elgar.

Sobel, R. S., Dutta, N., & Roy, S. (2010). Does cultural diversity increase the rate of entrepreneurship? *The Review of Austrian Economics, 23*(3), 269–286.

Somech, A., & Drach-Zahavy, A. (2013). Translating team creativity to innovation implementation the role of team composition and climate for innovation. *Journal of Management, 39*(3), 684–708.
Stam, E., & Wennberg, K. (2009). The roles of R&D in new firm growth. *Small Business Economics, 33*(1), 77–89.
Stenholm, P., Acs, Z. J., & Wuebker, R. (2013). Exploring country-level institutional arrangements on the rate and type of entrepreneurial activity. *Journal of Business Venturing, 28*(1), 176–193.
Tasselli, S., Kilduff, M., & Landis, B. (2018). Personality change: Implications for organizational behavior. *Academy of Management Annals, 12*(2), 467–493.
Tavassoli, S., Obschonka, M., & Audretsch, D. B. (2021). Entrepreneurship in cities. *Research Policy, 50*(7), 104255.
Teece, D. J. (2018). Profiting from innovation in the digital economy: Enabling technologies, standards, and licensing models in the wireless world. *Research Policy, 47*(8), 1367–1387.
The World Bank. (2022). *Government objectives: Benefits and risks of PPPs*. Accessed Sep 12, 2022, from https://ppp.worldbank.org/public-private-partnership/overview/ppp-objectives
Thite, M. (2011). Smart cities: Implications of urban planning for human resource development. *Human Resource Development International, 14*(5), 623–631.
Thuzar, M. (2011). Urbanization in Southeast Asia: Developing smart cities for the future? *Regional Outlook*, 96–100.
Ullah, F., Qayyum, S., Thaheem, M. J., Al-Turjman, F., & Sepasgozar, S. M. (2021). Risk management in sustainable smart cities governance: A TOE framework. *Technological Forecasting and Social Change, 167*, 120743.
UN. (2015). *Transforming our world: the 2030 agenda for sustainable development*. Accessed Sep 16, 2022, fromhttps://sustainabledevelopment.un.org/post2015/transformingourworld
Van Wijk, R., Jansen, J. J., & Lyles, M. A. (2008). Inter-and intraorganizational knowledge transfer: a meta-analytic review and assessment of its antecedents and consequences. *Journal of Management Studies, 45*(4), 830–853.
Verginer, L., & Riccaboni, M. (2021). Talent goes to global cities: The world network of scientists' mobility. *Research Policy, 50*(1), 104127. https://doi.org/10.1016/j.respol.2020.104127
Wales, W. J. (2016). Entrepreneurial orientation: A review and synthesis of promising research directions. *International Small Business Journal, 34*(1), 3–15.
Walshok, M. L., & Shragge, A. J. (2013). *Invention and reinvention: The evolution of San Diego's innovation economy*. Stanford University Press.
Wang, J., Lin, W., & Huang, Y.-H. (2010). A performance-oriented risk management framework for innovative R&D projects. *Technovation, 30*(2010), 601–611.
Washburn, D., Sindhu, U., Balaouras, S., Dines, R. A., Hayes, N., & Nelson, L. E. (2009). Helping CIOs understand "smart city" initiatives. *Growth, 17*, 1–17.
West, M. A. (1990). The social psychology of innovation in groups. In M. A. West & J. L. Farr (Eds.), *Innovation and creativity at work: Psychological and organizational strategies* (pp. 555–576). Wiley.
Willke, H. (2007). *Smart governance: Governing the global knowledge society*. Campus Verlag.
Wilson, F., Kickul, J., & Marlino, D. (2007). Gender, entrepreneurial self–efficacy, and entrepreneurial career intentions: Implications for entrepreneurship education. *Entrepreneurship Theory and Practice, 31*(3), 387–406.
Winters, J. V. (2010). Why are smart cities growing? Who moves and who stays. *Journal of Regional Science, 20*(10), 1–18.
Yigitcanlar, T., Foth, M., & Kamruzzaman, M. (2019). Towards post-anthropocentric cities: Reconceptualizing smart cities to evade urban ecocide. *Journal of Urban Technology, 26*(2), 147–152.
Yigitcanlar, T., Kamruzzaman, M., Buys, L., Ioppolo, G., Sabatini-Marques, J., da Costa, E. M., & Yun, J. J. (2018). Understanding 'smart cities': Intertwining development drivers with desired outcomes in a multidimensional framework. *Cities, 81*, 145–160.

Zacchia, P. (2018). Benefiting colleagues but not the city: Localized effects from the relocation of superstar inventors. *Research Policy, 47*(5), 992–1005.

Zait, A. (2017). Exploring the role of civilizational competences for smart cities' development. *Transforming Government: People, Process and Policy, 11*(3), 377–392.

Zheng, C., Yuan, J., Zhu, L., Zhang, Y., & Shao, Q. (2020). From digital to sustainable: A scientometric review of smart city literature between 1990 and 2019. *Journal of Cleaner Production, 258*, 120689.

Zimmermann, J., & Neyer, F. J. (2013). Do we become a different person when hitting the road? Personality development of sojourners. *Journal of Personality and Social Psychology, 105*(3), 515.

Zucker, L. G., Darby, M. R., & Brewer, M. B. (1998). Intellectual human capital and the birth of us biotechnology enterprises. *The American Economic Review, 88*, 290–306.

Chapter 3
Digital Platforms Enabling Long-Distance Knowledge Spillover in Smart Sustainable Cities: A Multilevel Framework

Andrea Ciacci

Abstract Digital platforms are information technologies where a great potential for innovativeness lies. By ensuring a constant connection among firms that put in common their knowledge and innovative processes, digital platforms are the foundations to create innovation able to raise the smartness and sustainability levels of cities. This chapter draws on the existing literature by aligning different theoretical insights. The main findings highlight that digital platforms operate in two ways to drive a firm's knowledge management. They enable knowledge exchanges with partners and facilitate the internal processes of knowledge management toward innovation. Local government can contribute to incentivizing the firm's participation in a digital platform by supporting business model innovation, information technologies-based capability development, and guiding the usage of innovative output within an urban environment. This innovation potentially contributes to satisfying social needs and attracting human capital which represents a relevant resource for the growth of the city.

Keywords Entrepreneurial ecosystem · Human capital · Digital platform · Knowledge spillover

3.1 Introduction

The objective of this chapter is to discuss how entrepreneurial ecosystems based on digital platforms manage knowledge to favor the growth of a smart sustainable city. Activities based on digital platforms allow partner firms to share knowledge. The deep interconnection characterizing the modern economic system claims new patterns of creation, sharing, and diffusion of knowledge. A firm participating in a digital platform can produce and offer services with higher potential, seeking social legitimacy derived from the public utility of its productive activities by leveraging incremental levels of knowledge. This chapter highlights that knowledge exchange can be nurtured by digital platform-based entrepreneurial ecosystems (Thomas et al., 2014). In this regard, digital platforms represent technologies enabling knowledge exchange (Eloranta & Turunen, 2016). Knowledge diffusion is a basilar element to

E. Ivaldi, A. Ciacci, *Smart Sustainable Cities and Knowledge-Based Economy*, Human Well-Being Research and Policy Making,
https://doi.org/10.1007/978-3-031-25038-5_3

grow a knowledge-based economy. Knowledge diffusion also benefits firms that build their competitive advantage on knowledge management and application (Ferraris et al., 2019). However, this chapter is not limited to knowledge management through digital platforms but embraces the issues of smart sustainable cities. This chapter argues that positive interactions between entrepreneurial actors and cities can occur when they share common interests. For instance, when a firm adopts a strategic orientation devoted to stakeholders and community well-being. Overall, this chapter aims to offer a framework to align the literature on a knowledge-based economy, entrepreneurship, and smart sustainable city by highlighting the positive impact that technology can have in terms of city growth. In this regard, the digital platform is the conceptual solution that bridges different streams of literature. Digital platforms are steadily acquiring notoriety in the business world (Annarelli et al., 2021; Cenamor et al., 2019; Jean & Kim, 2020; Shree et al., 2021). They have revolutionized many business models (Kohtamäki et al., 2019; Täuscher & Laudien, 2018). This tool brings benefits to businesses that adopt it effectively. But it can have a widespread effect. It can be the key to innovating by sharing and acquiring knowledge in a smarter way. The author argues that digital platforms can be the engine of a knowledge-based economy as well as a resource that businesses and public institutions must deploy to fuel growth at the city level through social needs satisfaction and human capital attractiveness.

Little is known about the properties an urban system should possess to achieve socially desirable objectives. The conceptual model proposed in this chapter shows that firms are pivotal actors within this system. They affect a city's social outcome by undertaking specific behaviors. For instance, networking attitude is one of the most relevant behaviors to allow knowledge can spill from a single organization to embrace a group of organizations. In the modern economy, networks manifest physically by geographical patterns or specific localization and virtually. Virtual networks of firms require technology assistance. For example, individual business units can share information, provide common standards, and ensure connectivity and IT capabilities through digital platforms (e.g., software or hardware connecting firms). All these factors positively influence the production, search, and delivery of digital content and goods among users of digital platform ecosystems (Karimi & Walter, 2015). Therefore, digital platforms are paramount business opportunities for firms. At the same time, they represent a serious challenge. Possessing advanced technological systems is not sufficient to participate in a digital platform. Technological systems must be supported by developing specific capabilities at the firm level, e.g., network and digital platform capabilities. The former is a firm's dynamic capability enabling the creation of interdependencies both within and outside the organization (Cenamor et al., 2019). Literature shows that network capabilities allow resource pool access, opportunities identification, quick responses to fast-changing market needs (e.g., urban community needs), innovation acceleration, transaction cost reduction, and gaining a better reputation (Lin & Lin, 2016). The latter indicates a firm's dynamic capability to integrate critical shared knowledge by leveraging internal resources and reconfiguring internal and external resources (Helfat & Raubitschek, 2018; Teece, 2018). Digital platform capability improves the firm's ability to communicate with other users' platforms (e.g., partners) and to acquire and organize structured information deriving from platform utilization (Cenamor et al., 2019). Both network and platform creation underlie complex systems of interrelationships. They refer to the creation of continuous knowledge exchange among business actors (e.g., platform users) and innovators (e.g., platform creators or leaders). However, if we extend the platform boundaries, we identify other influencing actors, such as government institutions (e.g., platform legal regulators) and urban communities (e.g., beneficiaries of knowledge spillover through platforms). The interaction of all these actors creates a multilevel synergy. It highlights that a digital platform-related issue is more than a solely technical or legal one. It involves long-term projects to align public decision-makers' intentions to the firm's convenience to create, innovate, and participate in a platform.

This chapter also provides insights into smart sustainable cities and the knowledge-based economy as concepts interrelated to entrepreneurship. The author argues that entrepreneurship is a valuable resource that smart sustainable cities can exploit to improve attractiveness in the eyes of people and workers. Entrepreneurship is a beneficial resource for knowledge-based economic growth. A city's growth is not to be intended from an economic-financial perspective exclusively. It also embraces sustainability, energy, transportation, people, education, health, well-being, safety, culture, ICT, welfare, and governance (Huovila et al., 2019; Penco et al., 2021). Therefore, the author refers to a broader conceptualization of the expression "city growth," in line with recent literature (Ciacci et al, 2021a, 2021b; Penco et al., 2020).

Overall, this chapter addresses a gap concerning smart sustainable cities under the knowledge-based economy paradigm. Few contributions emerged about the economy in smart sustainable cities and the interaction between a city type (e.g., smart and sustainable) and an economic perspective (e.g., knowledge-based economy). These concepts tended to grow distinctly. Their fast proliferation closed the time to unveil their intersection. This chapter aims to fill this gap by explaining the most relevant processes and dynamics.

Smart sustainable cities are places where technology must be functional to pursue the community's well-being (Ciacci et al., 2021a; Ivaldi et al., 2020; Musterd & Gritsai, 2013; Penco et al., 2021). In this regard, digital platform participation should be based on a social well-being improvement intent. Public decision-makers must advocate for the needs of their community to guide the platform's actions toward social goals. At the same time, they must provide adequate incentives to companies participating in risky projects because businesses are indispensable factors for innovation (De Massis et al., 2016). In their absence, the risk is to create a vacuum of innovative thinking.

3.2 Digital Platforms

A platform participated by a pool of firms represents the building blocks for developing complementary products, technologies, or services (Gawer, 2009). Digital platforms reorganize IT capabilities into software to integrate the transactions between different consumers and businesses and favor the interactions with the partners (De Reuver et al., 2018; Gawer, 2021; Srinivasan & Venkatraman, 2018).

For instance, a digital platform allows participating firms to extend the functionality of the products. They can do it by applying the basic features of a platform or by accessing its extensions (Karimi & Walter, 2015). Digital platform users interact with each other to find solutions that improve products and services. They invest trust, resources, and time to adhere platform that assumes a decisive role in their field. These efforts explain the shift from a value chain to a value network for value creation through a digital platform (Adner, 2017; Clarysse et al., 2014; Srinivasan & Venkatraman, 2018). The digital platform utility is due to the increase in quality of the product/service commercialized, the potential growth of the customer base, major cost structure efficiency, and more business opportunities and profits. The basic principle of digital platforms is to bring value to every user of the platform ecosystem (Adner, 2017; Clarysse et al., 2014; Jovanovic et al., 2021). A digital platform ensures a profit for the organization that created and maintains it[1] while also bringing advantages to the other participants (De Reuver et al., 2018).

It is possible to identify common elements of good digital platforms. They are the ease to use appeal for users, trustworthiness, security, and the promotion of knowledge exchanges between users. Their ecosystems are extendable by third parties. Finally, scalability is a core characteristic of all good digital platforms since it is necessary to ensure that changes in the platform do not compromise its functionality. Digital platforms enable firms to homogenize, edit, and distribute information on a large scale promptly (Cenamor et al., 2019; Yoo et al., 2010). Digital platforms have implications in terms of business model innovation, information management, and business value proposition (Cenamor et al., 2017). In addition, digital platform modifies the management of specific firms' resource. For example, immaterial and unstructured resources, such as big data, gain prominence in the digital platform era.

Firms can directly satisfy the customers' needs by matching supply and demand in the digital platform. In this way, firms engage in the process of value creation for themselves by product/service commercialization, and for customers by addressing their needs. In addition, firms can conveniently launch a platform to engage customers in the new product development process. Customers provide feedback and opinions to trigger the development process and product optimization. Platforms are flexible connecting hubs since they allow to save resources and time that otherwise should be devoted to administering surveys or conducting focus groups by obtaining real-time information (Sawhney et al., 2005).

[1]For some practical examples, check the following link: https://www.bmc.com/blogs/digital-platforms/#

Different types of digital platforms exist. The first distinction separates platforms to consumers (B2C platforms) (Sawhney et al., 2005) from platforms to business (B2B platforms) (Shree et al., 2021). The former provides information based on prior experiences, reviews, opinions, and knowledge needed by the local community (Karimi & Walter, 2015). The latter allows firms to store extensive local data, exchange information, and undertake other forms of collaboration. Much of this heritage is not accessible to the local community (Karimi & Walter, 2015). More in detail, we see platform applications every day. For example, we can refer to social media platforms (e.g., Facebook, Instagram, LinkedIn, Twitter), knowledge platforms (e.g., Yahoo! Answers, Quora, Stack Overflow), media sharing platforms (e.g., Spotify, Vimeo, YouTube), and service-oriented platforms (e.g., Airbnb, Grubhub, Uber). A second distinction focuses on the platform's sides in terms of users and customers (Evans & Schmalensee, 2016; OECD, 2019). They can be individuals or business organizations (OECD, 2019). The choice of the number of sides and their nature affects the pricing structure of the platform. For instance, the money side of a platform indicates the group of users that pays more than the marginal cost while the subsidy side refers to those that pay less than the marginal cost (Evans & Schmalensee, 2016; Gawer, 2021). A platform's sides impact the magnitude of the network effects and the diversified sources of revenue. Multiple sides can generate risks by creating an excess of complexity and conflicts between sides (Gawer, 2021).

Another aspect of primary importance concerns the digital interfaces of a platform (Gawer, 2021). They are structures shaping the modalities of interconnection between the multiple agents of a platform. Interfaces mark the boundary between agents, connect the multitude of data acquired, and act as a conduit for it (Gawer, 2021). A platform's interface shapes not only the design but more substantial aspects such as the degree of the platform control by the owner, the third parties' activities, complementary applications, and value capturing from innovations. The platform owner must decide on the platform set-up to balance the interests of the internal and external actors (Boudreau, 2017; Ghazawneh & Henfridsson, 2012; Nambisan et al., 2018; Yoo et al., 2012).

Cusumano et al. (2019) distinguish between transaction and innovation platforms. The former facilitates the exchange or transaction across sides, while the latter creates value by facilitating the innovation of goods and services. This chapter mainly refers to the *innovation platforms* (Cusumano et al., 2019; Gawer, 2021). Well-known examples of innovation platforms are Microsoft Windows, Google Android, Apple iOS, and Amazon Web Services. Despite these examples of successful large-size enterprises, innovation platforms are opportunities to enhance the performance of small- and medium-sized enterprises (SMEs) (Ben Arfi & Hikkerova, 2021; Cenamor et al., 2019; Nambisan et al., 2018). It is particularly relevant from a local perspective since business agglomerations involve many SMEs that represent an essential part of every entrepreneurial local fabric (European Commission, 2019; Florida et al., 2020).

From the theoretical standpoint, this chapter addresses the *platform ecosystem stream* (Thomas et al., 2014) based on the conceptualization of a platform as a touchpoint or hub within a technology-based business system. It is characterized by the intensive usage of information technology and the Internet, a multisided configuration, and high modularity. They serve as a coordination structure of a network of businesses by organizing assets, services, and technologies (Thomas et al., 2014). Thomas et al. (2014) identify the platform ecosystem as the hub or central point of control for a business system based on technology. The agents in platform ecosystems share and use common knowledge. By leveraging unique resources, a firm can integrate knowledge deriving by the platform to create new products, services, and complementary modules (Jovanovic et al., 2021). Digital platforms are facilitators of interactions (Boudreau, 2010) that contribute to information sharing, knowledge creation, processes, components, people, and relationships at the heart of many business models (Gawer, 2009; Robertson & Ulrich, 1998). The users of a digital platform together form an *ecosystem*, i.e., “the alignment structure of the multilateral set of partners that need to interact for a focal value proposition to materialize” (Adner, 2017, p. 40).

A conversion toward a pervasive digital mindset is required to lead or participate in a platform. It is the premise for every adoption of digital tools can effectively work (Akter et al., 2016; Ben Arfi & Hikkerova, 2021; Grover et al., 2018; Reis et al., 2020). Together with a platform-oriented culture, a firm must develop specific capabilities surrounding the execution of the platform-based activities over time (Helfat & Raubitschek, 2018; Teece, 2017). These capabilities vary depending on the stage of development of a platform. They range from sensing and planning capabilities during the earlier stages of a platform lifecycle to transformative and innovative capabilities when the platform achieves a more advanced level of maturation (Teece, 2017).

3.3 Knowledge Management in the Digital Platforms

Knowledge flows through the digital platform (Bouncken et al., 2021). This process works inside and outside a firm since it can receive knowledge from partners and share its own knowledge through the platform. The firm purifies knowledge and selects it strategically. In this phase, the key to success concerns the ability of the firm to align the newly acquired knowledge with the knowledge possessed. This involves a succession of internal learning processes and selection prior to alignment (Argote & Miron-Spektor, 2011; Eisenhardt & Sull, 2001; Sirén & Kohtamäki, 2016; Sirén et al., 2012). Knowledge management interrelated to digital platform-based activities imply that the information or knowledge acquired from a platform is treated internally by an organization. Specifically, a process toward innovation based on a platform works at different stages embracing the acquisition of knowledge, organizational learning, and knowledge transmission between departments and team groups. These actions allow a firm to build on the knowledge flowing through the platform to fuel the internal innovation processes (Ben Arfi & Hikkerova, 2021; Nonaka, 1994). We could summarize this step as a sequence of internally coordinated efforts to capitalize on the inbound knowledge. Ben Arfi and Hikkerova (2021) highlight that a firm can use a digital platform to share knowledge internally. The success of this process depends on the occurrence of specific conditions, such as the promotion of participatory workplaces where members are integrated into the innovation project. The leadership style is also crucial. Transformative styles are more appropriate to support innovation projects. Finally, motivation is decisive in incentivizing collective innovation-oriented actions. It is conceived as a set of intangible elements, such as commitment, trust, care, and other shared values facilitated by socialization. These conditions together contribute to gaining high-quality tacit knowledge sharing and transfer within an organization.

The kind of knowledge that partners exchange through a platform can involve the design of a product or service, know-how, technology applications, limitations, production techniques, mathematical models, and testing methods (Boudreau, 2012; Gawer & Cusumano, 2014; Robertson & Ulrich, 1998) depending on the platform's specialization and the industry of reference.

Recent studies demonstrate that firms in servitization frequently use digital platforms (Cenamor et al., 2019; Eloranta & Turunen, 2016). In this stream, digital platforms are considered a key driver to offer advanced services by leveraging valuable information (Baines & Lightfoot, 2014; Cenamor et al., 2017; Opresnik & Taisch, 2015). Firms can use digital platforms to share cumulative knowledge. This activity relies on the network's knowledge exploitation. The platform's partners engage in a value co-creation mechanism (Thomas et al., 2014) by trading the information possessed by a firm for valuable new knowledge owned by other partners. Partners use the platform to innovatively combine knowledge-based resources to reach the best solution for their customers (Eloranta & Turunen, 2016). Therefore, digital platforms call for the importance of managing complex networks of partners (Boudreau, 2010). The success of a platform-based business model depends on the right management of these intricate relationships (Cenamor et al., 2019).

Therefore, the field of application of digital platforms is vast. It involves both processes of inbound and outbound knowledge sharing and transformation. The firms in digital platform ecosystems share and use common resources and knowledge to create new complementary modules. The digital platforms work as facilitators of interactions. They have contributed to placing knowledge-based processes at the center of many business models (McAfee et al., 2012; Kohtamäki et al., 2019; Van Alstyne et al., 2016). Platform integration improves internal communication and coordination among teams, departments, and business units. The digital platform represents an integrative architecture for the centralization and formalization of internal information flows (Dominguez Gonzalez & Massaroli de Melo, 2018; Cenamor et al., 2019; Helfat & Raubitschek, 2018). Therefore, digital platforms facilitate the internal communication and coordination of resources, capabilities, competencies, activities, goals, and knowledge (Helfat & Campo-Rembado, 2016). Digital platforms also enhance relational skills and knowledge exchange with partners. The participants manage to change networks of partners by leveraging the modular architecture that characterizes the digital platform (Baldwin, 2012;

Marion et al., 2015). In summary, digital platforms allow firms to improve their ability to share information, knowledge, ideas, assets, resources, and processes by communicating with external partners and acquiring and organizing structured information and knowledge from external partners (Cenamor et al., 2019). The field of application of digital platforms is both external and internal to organizational boundaries.

From a conceptual standpoint, digital platforms represent a tool in which a knowledge-based economy should invest its future. A knowledge-based economy is "directly based on the production, distribution, and use of knowledge and information" (OECD, 1996, p. 3). The literature highlights the pivotal role of knowledge production toward innovation (Bell, 1973; Romer, 1990), productive process improvement (Kochan & Barley, 1999), and collaboration between different organizations (Drucker, 1993; Nonaka & Takeuchi, 1995). Other studies have emphasized the importance of knowledge-based activities for wealth and development at the city level (Ivaldi et al., 2020). Therefore, digital platforms seem to have the potential to assist the growth of a smart sustainable city by benefiting its essential components.

Information technologies play a central role in knowledge management. These systems allow to store codified knowledge, favor the interaction among businesses, and connect the professionals' tacit knowledge (Alavi & Leidner, 2001; Davenport et al., 1998). Specifically, the ability to integrate and manage different knowledge-based resources is considered a significant predictor of the likelihood of innovations (Schumpeter, 1934). A way to gain competitive advantage consists in effectively applying knowledge in new contexts.

Knowledge-based smart sustainable cities are founded on knowledge agglomerations (Ivaldi et al., 2020). In a knowledge-based economy, urban agglomerations are "knowledge agglomerations" (Ivaldi et al., 2020, p. 47; Penco, 2015, p. 822). Business proximity and the formation of clusters at a local level remains a relevant issue (Amezcua et al., 2020; Florida et al., 2008; Florida, 2002; Qian & Acs, 2013; Tavassoli et al., 2021). The growth of such agglomerations must improve in density, diversity, and specialization since they can provide a base to innovate (Andersson et al., 2016). Literature on local clusters of firms unveiled their prominent role in knowledge sharing, proliferation, and transmission (Anselin et al., 1997, 2000; Jaffe et al., 1993; Peri, 2005; Qian & Acs, 2013). Positive contagious effects descend from the presence of firms' agglomerations. If physical clusters represent a key to managing inbound and outbound knowledge flows, information technologies constitute a foundation to manage knowledge digitally. ICT development has provided many tools to spread long-distance knowledge. Digital platforms represent one of these. By definition, knowledge is immaterial, equated with information (Barley et al., 2018). An important distinction concerning the nature of knowledge is between explicit (codified and material) knowledge and tacit (intangible and context-specific) knowledge. Explicit knowledge can be externalized through symbols, objects, and language (Barley et al., 2018). Tacit knowledge can also refer to people's knowledge and its application in action (Polanyi, 1966). To wit, explicit knowledge is transferred during exchange among the platform's business users.

Tacit knowledge lies in the creativity of the people. It is decisive in the new business formation and characterizes high-skilled sought-after workers.

To the knowledge exchanges based on physical proximity, new modes for long-distance knowledge exchange are coming to the fore (Täuscher & Laudien, 2018). New ways to share knowledge take place in a virtual environment. This way of knowledge management does not deny or preclude the spatial approach, where geographical configurations of firms' physical distribution in a city acquire prominence (Porter, 1998). They integrate each other, ensuring continuity in knowledge flowing over time and beyond rigid spatial patterns. A conceptual model where a knowledge transfer system in a virtual environment integrates the ordinary physical agglomeration remarks the importance of having many firms physically engaged in innovative activities in the urban space. In fact, the creation of social value in a city can happen when firms have an interest (or convenience) to undertake activities beneficial to society. In a further step, when their plans come to fruition, innovative firms spill the innovations within the community, generating social value. Firms operating in the urban territory commit to the social promotion of that same territory through innovation grounding.

In today's world, knowledge must flow over the distance. Fast-changing times require rapid and smart solutions. Supply chain actors must continuously solve problems, perfect their activities, and organize projects for innovation. Platform-based solutions allow firms to manage their activities in a smarter mode. A platform-based innovation system can help to overcome contingencies, ensuring operational continuity for every organization (Soluk et al., 2021; Tønnessen et al., 2021).

In addition, knowledge is a critical factor for innovation (Andersson et al., 2016; Papa et al., 2018; Santoro et al., 2020). Nonaka (1994) consider the innovation process as an ongoing dialog between tacit and explicit knowledge. Therefore, a knowledge-based economy depends on the capacity to use the knowledge capital at hand to innovate. Knowledge management mechanism underlies the promotion of innovation. The knowledge-based economy seems to be an innovation-oriented economy. Better, the growth of smart sustainable cities based on a knowledge-based economy seems to depend on the innovative firms' ability to capitalize on their knowledge, i.e., building upon their knowledge management systems to acquire, transform, integrate, and create new knowledge.

3.4 Product Development, Environmental Technologies, and Innovation Enhancing the Smartness and Sustainability of a City

The scope of smart sustainable cities encompasses a complex balance between environmental sustainability and infrastructures (Höjer & Wangel, 2015) (e.g., information and smart technologies, urban metabolism). A smart sustainable city "meets the needs of its present inhabitants, without compromising the ability for other people or future generations to meet their needs, and thus, does not exceed local or planetary environmental limitations, and where this is supported by ICT" (Höjer & Wangel, 2015, p. 342). Crucial issues of smart sustainable cities refer to the quality of life, services, and environmental sustainability improvement (Garau & Pavan, 2018; Ivaldi et al., 2020). Livability is the tension that shapes smart sustainable city planning.

Concerning innovation in a smart sustainable city, an intrinsic complexity emerges. Innovation refers to a wide range of products or services. Schumpeter (1934) defined innovation as the first introduction of a new product, process, or system into the economic and social system. The innovation is not limited to the previous outcomes but involves organizational and managerial innovations, the opening of new markets, the discovery of new sources of supply, and finance. Different types of innovation match the theoretical framework presented in this chapter. All the innovations are based on the intensive usage of knowledge during the transformative processes or as an outcome. Such as innovation must conform to the smart and sustainable identity of the city. The application of this framework ranges from environmental technologies to services improving the citizens' quality of life. The former indicates "production equipment, methods and procedures, product designs, and product delivery mechanisms that conserve energy and natural resources, minimize an environmental load of human activities, and protect the natural environment" (Shrivastava, 1995, p. 185). They fit the smart sustainable city's vision. Environmental technologies facilitate the city's transition toward a more sustainable model (Elliot, 2011). At the same time, they provide many opportunities for businesses to collaborate and express their innovativeness (Shrivastava, 1995). Examples of environmental technologies gaining momentum are solar panels and wind turbines (renewable energy), sensors, intelligent lighting systems, electric vehicles, and Direct Air Capture systems.[2] Among the multiple fields of applications, we recognize pollution control and product lifecycle impact reduction (Huang et al., 2016; Oltra & Saint Jean, 2009; Zackrisson et al., 2010). Therefore, they distinguish a high degree of utility in a smart sustainable city oriented to livability improvement and environmental sustainability. These products require high technical skills, foresight planning, knowledge acquisition, and creation systems. Drawing on platforms distinguishing characteristics, the author argues that platforms constitute the touchpoint to ensure the innovation process succeeds.

At the global level, the Partnership for Environment and Disaster Risk Reduction (PEDRR)[3] has been founded. It is a global alliance of UN agencies, NGOs, and specialist institutes. The PEDRR aims to stimulate the implementation of Ecosystem-based Disaster Risk Reduction (Eco-DRR), where partners collectively implement planned initiatives and activities. The PEDRR is more than a knowledge

---

[2]For more information see: https://edinburghsensors.com/news-and-events/impact-of-technology-on-the-environment-and-environmental-technology/

[3]For more information, see: https://pedrr.org/about-us

exchange network. It is a global advocate for increased investments in ecosystem-based approaches to reducing disaster and climate risks.

One could imagine a re-proposition of a platform alliance in which participants, under institutional auspices, envision the development of new environmental technology products and their application at the city level. Cities and their inhabitants would benefit from such innovations driven by the innovative creativity of companies. This example is only a nod to the myriad applications of innovations based on knowledge transmission and integration through digital platforms. The factual reality speaks of a system that is still under construction. The key to launching regular productive collaborations lies in the incentives provided to innovative firms to promote complex business model systems based on digital platforms that drive business innovativeness while limiting economic and financial risks. Innovative firms may find themselves short of incentives to implement disruptive new technologies. But when it comes to downstream innovations, i.e., applications that build on the primary systems of innovation, it has been shown that there may be a case for innovation, given the high profitability of such firms. Thus, digital platforms would constitute a high-return ecosystem for such firms. On the side of local governments, the main issues are related to the deployment of such products, the transition to smarter sustainable city models, and the monitoring of progress on a periodic basis. Public institutions could even decide to join the digital platforms and collaborate from within with private parties by sharing complementary goals, knowledge, and expertise. This collaboration would facilitate the leadership role of the institutions to adopt innovations.

Beyond any other example, literature on digital platforms looks at innovation with particular interest (Gawer, 2021; Nambisan et al., 2019). Nambisan et al. (2019) revealed that three characteristics of digital technologies (i.e., affordances, generativity, openness) contribute to the search for innovation by enabling flows of knowledge during product development stages, leveraging the "multipart representation" (p. 5), and expanding the set of possible innovative actions. Jovanovic et al. (2021) highlight that innovation mechanisms characterize the platform archetypes and directly contribute to the overall platform value increase. Gawer (2021) refers to the so-called innovative platforms to indicate the specific orientation undertaken by the platform's participants. Ben Arfi and Hikkerova (2021) leverage the concepts of corporate entrepreneurship, product innovation, and knowledge conversion to show that digital platforms enable a favorable climate where knowledge management mechanisms are improved. Under these conditions, corporate entrepreneurship positively affects product innovation.

Therefore, the literature demonstrates that specific kinds of digital platforms facilitate the process of innovation. This aspect comes to the fore in the smart sustainable city context, where individual and community growth depends on the innovativeness that characterizes the urban ecosystem (Penco et al., 2020).

3.5 Smart Sustainable Cities Based on Knowledge Economy Objectives

A smart sustainable city is supported by a pervasive presence and massive use of advanced ICT connecting various urban domains and systems, which enable the city to become more sustainable and to provide citizens with a better quality of life (Bibri & Krogstie, 2017).

The smart sustainable prototype city integrates the different dimensions of sustainability into the traditional smart city (Zheng et al., 2020). The smart sustainable city embodies the high-standing features of quality of life or well-being (Hara et al., 2016; Pinna et al., 2017). High quality of life can contribute to attracting and retaining skilled knowledge workers in the city (Penco, 2015). Quality of life is the driver to expanding the knowledge base of a city. A knowledge-based economy manifests when many economic actors (e.g., entrepreneurs) organize the latent knowledge into a structured business (Qian & Acs, 2013; Tavassoli et al., 2021).

A smart sustainable city involves environmental and functional elements, such as infrastructures, ICTs, technologies, and sewage, water, energy, and waste systems management (Belanche et al., 2016; Piro et al., 2014; Shin, 2009). These factors make the city more livable (Garau & Pavan, 2018).

The essential vision of the smart sustainable city embodies the satisfaction of its inhabitants' needs to enhance the quality of life in place (Bibri & Krogstie, 2017; Khan et al., 2015; Marsal-Llacuna et al., 2015; Batty, 2012). The positive impacts of the smart sustainable city on quality of life generate a higher attractiveness toward the external. The accumulation of rich human capital and knowledge potential is decisive for (economic-financial) growth at the city level (Thite, 2011; Glaeser et al., 1995). These interrelated mechanisms define the city's basic principles and identity.

Businesses are essential contributors to the challenge of environmental degradation (Elliot, 2011). In addition, modern businesses intensively use technology to pursue ordinary and non-ordinary activities. Entrepreneurship and smart sustainable cities seem to have a lot in common. Entrepreneurship is a bearer of the instances of a smart sustainable city in innovation.

Business agglomerations leveraging platforms to share and acquire knowledge reflect the salient traits of a city model based on technological centrality and a knowledge-based economy. For this reason, the author argues the theoretical-conceptual complementarity between a platform-based network and a smart sustainable city.

In this framework, a smart sustainable city can indirectly benefit from the technology aid through the action of the firms. Firms canalize flows of knowledge acquired by the digital platform to achieve innovation. Firms can develop innovations with the potential to satisfy the city's needs by using digital platforms. The knowledge economy is a base of knowledge-intensive productions and services contributing to speeding up technological and scientific development (Powell & Snellman, 2004).

3.6 Conceptual Framework

Based on the previous conceptualizations, this paragraph summarizes the complete framework (Fig. 3.1). It is multilevel, involving firms, networks, local institutions, and the city, and highlights how different elements interact to pursue growth at the city level.

A firm must develop dynamic capabilities to adopt new technologies or undertake a business model based on a digital platform. Capabilities are the winning secret of a business. They assist the business model innovation necessary to align the business processes with the new technologies adopted (Ciampi et al., 2021; Heider et al., 2021; Randhawa et al., 2021; Teece, 2018). Dynamic capabilities are inimitable and unique and represent the substrate where all internal innovation rests. In this context, dynamic capabilities bridge the old and new ways of doing business. However, the dynamic capabilities development process and business model innovation are expensive, tough to support over time, and uncertain. For the firms can be hard sustaining this effort. Specifically, SMEs are characterized by liabilities of smallness and newness (Miller et al., 2021). Governmental incentives can help a firm to succeed in the expensive process of capability development and business model innovation. Not only governments at various levels can bestow economic resources that relieve firms of complete and utter risk-taking, but they can instill the consciousness that the firms are not alone. In their perception, it is like they count on the support of an authority to receive an additional positive boost. This can have further implications concerning the perception of the business environment (Ciacci et al., 2021b), attracting new pools of businesses. The institutions' proximity to the private sector can benefit the entire entrepreneurial ecosystem (Teece, 2018). Firms may also need specialized managers for innovation (Clarysse et al., 2014). In fact, the digitalization process implies an organization develops a new mindset. For a firm, being able to rely on a skilled professional to play a leading role and guide this transition is a potential element of success.

Once the firm has created the conditions, it dives into the digital platform world, where knowledge exchanges with partners take place. The firm must integrate what emerges during interactions with partners into its business strategy. In this regard, the ability to learn (Sirén & Kohtamäki, 2016) and align new insights with the knowledge already possessed become decisive to avoid information overload, a precursor of strategic failure. Digital platform activity is a knowledge-leveraging play for each participant.

The firm achieves its productive scope based on platform activity and internal reconfigurations. As written before, this framework particularly fits environmental technology-oriented and service firms. In this case, local governments can drive production to satisfy social needs. Local governments are aware of the needs of urban areas. They can enable efficient use of the innovations produced. Firms also should be at their disposal to collaborate with local governments since their knowledge of the territory can offer a secure outlet and a proper location for firms' products and services.

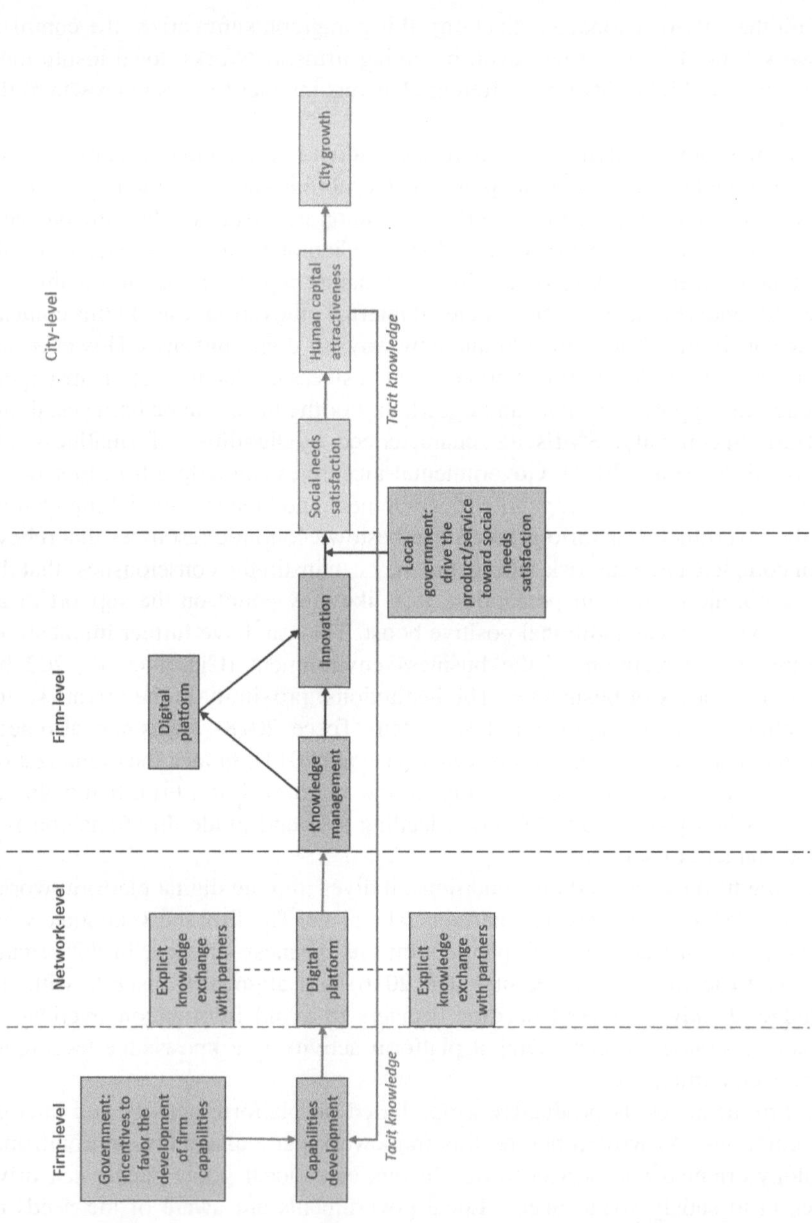

Fig. 3.1 Conceptual framework

Products and services employed for social purposes increase the community's well-being. Such an approach is ideal for designing a smart sustainable city where smartness is aimed at enhancing sustainability and social well-being. A well-perceived place is more likely to attract new people (e.g., highly skilled employees, young university students). People contribute to strengthening the city's human capital (Moos et al., 2019). They enjoy the city due to its commodity, specifically if it has a modern infrastructure system (Glaeser & Gottlieb, 2006; Turok, 2008). New people are useful to the city. They represent a potential source of knowledge from which a firm can draw. They are directly engaged in the city's growth. In this way, a bidirectional relationship establishes between the city's people and entrepreneurial activities devoted to the city's growth.

The adoption of a digital platform is a complex process. A firm that wants to develop a platform-based business model needs time and money to acquire the technical architecture, train employees, and recruit already skilled employees. Many reconfigurations can occur before the firm achieves the desired result. It can be also more difficult for SMEs characterized by liabilities, smallness, newness, and, more in general, scarcity of resources (Eggers, 2020). SMEs' problems are also the city's problems. Many firms composing business agglomerations are SMEs (Eggers, 2020). They can store great innovativeness. Acs and Audretsch (1987, 1988) demonstrated that a strong innovation spirit lies in SMEs. In many cases, they could struggle to implement platform-based functions inside their business without success. They could need platform-based activities to capitalize on benefits deriving from network effects. Alternatively, they could not have the possibility to turn to a platform-based model of business on their own. They could need external support. In this regard, the role of local government is crucial. The success of a transition toward a platform-based business could depend on government intervention. Private-public partnerships offer substantial public benefits (e.g., improved service quality, risk-sharing with the private sector, and cost savings) (Bloomfield, 2006). These partnerships increase the capacity of the government to deliver infrastructures, services, and products through risk-sharing agreements. Innovative public-private partnerships provide mutual benefits to the parties involved (Emerson et al., 2012).

To summarize, it is possible to formulate some propositions:

1. The conversion toward a digital platform-based business model implies a radical transformation of a firm's business model that calls for governmental support if the firm pursues an activity of public interest with a widespread impact. The digital platform intervenes in two ways to drive a firm's knowledge management, i.e., by allowing knowledge exchanges with partners and facilitating the internal processes of knowledge management toward innovation.
2. Local government contributes to guiding the innovation to the application in the urban environment.
3. The innovation generated by digital platforms creates a positive vortex by satisfying social needs and attracting human capital that has beneficial effects on the city's growth.

4. This sequence of development restarts thanks to the contribution offered by the human capital as a reservoir of tacit knowledge that represents the pool of new creative minds for the firms.

However, firms and local governments must face further challenges to create the best conditions for growth. On the one hand, it emerges a problem of profitability for firms. On the other hand, it emerges a problem concerning the actual impact of the innovation on the urban community that local decision-makers must handle.

Engaging in innovative activities is expensive. Innovation involves tendentially high investments under risk and uncertainty. Innovative firms must consider that the risk of failure is an unavoidable component of innovation (Hughes et al., 2018). The search for new disruptive ideas involves that much time could be devoted to experimental activities. Strong incentives should characterize the field of innovation to stimulate firms in perpetrating their actions over time.

Teece (2018, p. 1383) sentences that "value capture should be part of every exercise in strategy, business model design, and innovation. Merely being the pioneer is not the road to riches. It never has been." Firms should support their technical architecture with an adequate commercial strategy. However, creating the best internal conditions to innovate is not sufficient to ensure high returns. Business models based on licensing are difficult to implement from a value-capture perspective. They could reveal unprofitable and discourage future activities of innovation. Policymakers should treat these activities as the result of R&D enabling new downstream economic activity. Therefore, rewards should be calibrated to sustain the entire expensive innovation cycle by recognizing incentives for innovations with a potentially high impact (Teece, 2018). Profitability and regulatory issues affect the linearity of this process toward success. Cities' managers should consider that, without incentives and economic assistance, innovative firms may be reluctant to engage in risky activity. In this regard, Teece (2018) states: "Policy makers must recognize the challenges that innovators face with respect to capturing enough value to continue innovating in the future. The problems are amplified for inventors of enabling and general-purpose technologies. To keep a society's innovation engine fueled, the government needs to judiciously support value capture, not just value creation. Otherwise, incentives to innovate will be compromised and/or the government itself will need to fund enabling technologies at levels not hitherto contemplated."

The second problem concerns how institutions can leverage inventions to maximize their impact on society. Local governments should conceive the private efforts as an opportunity for growth for the city itself. Incentivizing firms to remunerate their efforts stimulates the firms to perpetuate these behaviors over time. The creation of this system would benefit society. To adopt a systemic perspective, governments should collaborate with privates by establishing joint development plans. The two parts should interact from the initial step of this process to set fair and favorable terms of the agreement. By stipulating contracts, the parts should indicate the ways and context of using inventions to recreate in advance the best conditions for this to happen regularly and for the invention to have a widespread impact.

The success in commercializing an innovative product or service largely depends on the degree of utility attributed to the product or service. If the product is characterized by high usability or a natural inclination of usage in day-to-day activities focused on the well-being of individual units in a community, the likelihood that the product will be sold increases significantly. If the attention unit is a smart city, the usefulness of the product is greater when it helps meet one of the city's primary goals. Since the natural goal of a smart city is to make urban environments more livable with the help of technology, major innovations should be aimed at that goal.

Technology is central to joint city-entrepreneurship development. As in the case of a firm in which platform capability powers an ecosystem, a city, i.e., an ecosystem of entrepreneurial agglomerates, might reinforce its platform capability to achieve growth at the ecosystem level. Technology is pivotal in providing operational support allowing platform activities. Platforms allow firms to share best practices, lower costs, gain reputation, and address customers and market needs. Digital platforms especially enhance the knowledge-sharing and management processes between partner firms and within single firms.

3.7 Conclusions

The field of application of this conceptual framework concerns the cities where technology and sustainability have a pivotal role. These cities undertake actions to stimulate innovation. Specifically, cities with a knowledge-based economy can benefit from this system adoption since it creates the architecture to spill over knowledge effectively. In summary, this conceptual model is applicable to knowledge-based smart sustainable cities or those cities that are working to build such as economy type. Since this conceptual alignment among different streams is relatively new, future research could involve it in a more precise theory by providing examples, if any, of such digital platform-based collaborations. This chapter also highlights two problems that urban governments should consider. The first problem is related to the firm's convenience to embark on innovative activities. The second refers to innovation usage to satisfy a local community's needs. These relevant issues should be exhaustively addressed in future research.

The conceptual framework presented in this paper shows how a firm's functions serve to fulfill the city's growth. This paper aims to pursue a general alignment of different research streams (e.g., smart sustainable city, knowledge-based economy, digital platform, private and public policies, human capital, well-being, and quality of life). Specifically, it shows that knowledge is transferred from one firm to another via digital platforms. Digital platforms work as facilitators of interactions and touchpoints of multiple knowledge sources. Firms must develop capabilities to adopt a digital platform business model.

This chapter shows that the city's growth process based on digital platforms is multilevel. Therefore, implications arise at the firm, network, institutional, and community levels. The number of actors involved is high. It implies that the level of intrinsic complexity is high. It is difficult to trace these processes univocally to a linear sequence. It represents the main limitation of the study. The conceptual framework already demonstrates that government intervention can positively alter the course of events. This point addresses a claim for future research, i.e., studying the micro-processes working in the framework at multiple levels.

The chapter deals with the management of knowledge inside and outside the firm. This framework recommends that digital platform participation is vain if not sustained by the apposite firm's capability. Strategic planning must underpin capability development by ensuring investments, time, and adequate training (or adequately trained employees). Knowledge management implies multiple strategy adaptations over time through the iterative processes involved in strategic learning. This framework emphasizes the process of value creation both for firms and the city, as well as its inhabitants. The author hopes this framework is helpful to explain to city managers how collaborating with firms is beneficial to their city. Private-public partnerships should be frequently undersigned to intervene in multiple circumstances.

By definition, a cluster involves institutions within firms' agglomerates (Porter, 1998). Being a platform a form of clusterization, local governments should understand the potential of a digital platform and its modes of usage. Public institutions should enter the world of digital platforms to collaborate with participating businesses and drive the use of end products. This participative approach would help firms to understand how to employ their products toward the city's growth. Finally, firms participating in a digital platform are likely to co-create value (Clarysse et al., 2014). However, co-creation does not mean co-capturing (Clarysse et al., 2014). The risk of low-expected returns for firms is high. For this reason, they need facilitation. Different approaches may come to the fore to ensure support: Business angels, accelerator activities, and cross-border venture capital provisions involving both cross-national and cross-regional borders are some examples (Clarysse et al., 2014). Future research could focus on analyzing the effects of crowdsourcing as a possible source of financing. Crowdfunding can influence both the processes and outcomes of a digital platform activity (Nambisan et al., 2019). Therefore, one may argue that this financing approach can lead to a digital platform-based model in which society has a prominent role.

Sydney
Country: Australia
Inhabitants: 5,231,147
Density: 433/km^2 (1120/sq. mi)
Total area: 12,367.7 km^2 (4775.2 sq. mi)

(continued)

The transformation of Sydney, the capital of the Australian state of New South Wales, into a real Smart City is driven primarily by technological innovation. In order to maintain its reputation as a world-leading city for citizens, businesses, and tourists and to cope with its growing population, at the beginning of 2020, the local administration presented the Smart City plan, drafted jointly with stakeholders, citizens, and city staff. Since the publication of this document, the administration has received a lot of feedback and comments, mostly on the subjects of innovation, data, and economic development, which have since been integrated into a later draft. The city of Sydney's strategy to become a Smart City is articulated in five key points.

In a digitally evolving world, it is necessary to allow everyone free access to technological tools: The first goal thus concerns the creation of a digitally aware and resilient to change community. Secondly, Sydney intends to maintain its role as a leading city in economic development by attracting professionals from around the world able to accelerate the process of urban innovation. To this end, the administration intends to invest in innovative ecosystems through the creation of a collaborative environment open to the exchange of ideas, the training of workers, and making the tourist experience as immersive as possible. The third goal of Sydney is to be resilient to future challenges, including climate change. This target can be met by integrating concepts from the circular economy and sustainable resource management, as well as by continuously monitoring activities and gathering data.

The establishment of inclusive and accessible urban environments using digital technology is the Smart City Plan's fourth stated goal, while the provision of specialized services tailored to residents' actual needs is its fifth goal (e.g., through the creation of a platform putting the municipality in direct contact with citizens).

Although the City of Sydney's program was just announced in 2020, the local administration has already been working on projects linked to smart cities (such as an open data portal, smart traffic lights, smart lighting, sustainable waste management, etc.) for several years. One of the focuses of the Australian metropolis's strategy is the community's technological inclusion: The aim is to replace antiquated teaching techniques with advanced technological instruments able to offer everyone constant training and updating of skills. Finally, in economic terms, Sydney plans to adopt the "innovation economy" model and boost its GDP per capita, which is already among the highest in the world.

References

Acs, Z. J., & Audretsch, D. B. (1987). Innovation, market structure, and firm size. *Review of Economics and Statistics, 69*(4), 567–574.

Acs, Z. J., & Audretsch, D. B. (1988). Innovation in large and small firms: An empirical analysis. *American Economic Review, 78*(4), 678–690.

Adner, R. (2017). Ecosystem as structure: An actionable construct for strategy. *Journal of Management, 43*(1), 39–58.

Akter, S., Wamba, S. F., Gunasekaran, A., Dubey, R., & Childe, S. J. (2016). How to improve firm performance using big data analytics capability and business strategy alignment? *International Journal of Production Economics, 182*, 113–131.

Alavi, M., & Leidner, D. E. (2001). Review: Knowledge management and knowledge management systems: Conceptual foundations and research issues. *MIS Quarterly, 25*(1), 107–136.

Amezcua, A., Ratinho, T., Plummer, L. A., & Jayamohan, P. (2020). Organizational sponsorship and the economics of place: How regional urbanization and localization shape incubator outcomes. *Journal of Business Venturing, 35*(4), 105967.

Andersson, U., Dasí, Á., Mudambi, R., & Pedersen, T. (2016). Technology, innovation and knowledge: The importance of ideas and international connectivity. *Journal of World Business, 51*(1), 153–162.

Annarelli, A., Battistella, C., Nonino, F., Parida, V., & Pessot, E. (2021). Literature review on digitalization capabilities: Co-citation analysis of antecedents, conceptualization and consequences. *Technological Forecasting and Social Change, 166*, 120635.

Anselin, L., Varga, A., & Acs, Z. (1997). Local geographic spillovers between university research and high technology innovations. *Journal of Urban Economics, 42*(3), 422–448.

Anselin, L., Varga, A., & Acs, Z. (2000). Geographic and sectoral characteristics of academic knowledge externalities. *Papers in Regional Science, 79*, 435–443.

Argote, L., & Miron-Spektor, E. (2011). Organizational learning: From experience to knowledge. *Organization Science, 22*(5), 1123–1137.

Baines, T., & Lightfoot, H. W. (2014). Servitization of the manufacturing firm: Exploring the operations practices and technologies that deliver advanced services. *International Journal of Operations and Production Management, 34*, 2–35.

Baldwin, C. Y. (2012). Organization design for business ecosystems. *Journal of Organization Design, 1*(1), 20.

Barley, W. C., Treem, J. W., & Kuhn, T. (2018). Valuing multiple trajectories of knowledge: A critical review and agenda for knowledge management research. *Academy of Management Annals, 12*(1), 278–317.

Batty, M. (2012). Smart cities, big data. *Environment and Planning B: Planning and Design, 39*, 191–193.

Belanche, D., Casaló, L., & Orús, C. (2016). City attachment and use of urban services: Benefits for smart cities. *Cities, 50*, 75–81.

Bell, D. (1973). *The coming of post-industrial society*. Basic Books.

Ben Arfi, W., & Hikkerova, L. (2021). Corporate entrepreneurship, product innovation, and knowledge conversion: The role of digital platforms. *Small Business Economics, 56*(3), 1191–1204.

Bibri, S. E., & Krogstie, J. (2017). Smart sustainable cities of the future: An extensive interdisciplinary literature review. *Sustainable Cities and Society, 31*, 183–212.

Bloomfield, P. (2006). The challenging business of long-term public–private partnerships: Reflections on local experience. *Public Administration Review, 66*(3), 400–411.

Boudreau, K. (2010). Open platform strategies and innovation: Granting access vs. devolving control. *Management Science, 56*(10), 1849–1872.

Boudreau, K. J. (2012). Let a thousand flowers bloom? An early look at large numbers of software app developers and patterns of innovation. *Organization Science, 23*(5), 1409–1427.

Boudreau, K. J. (2017). Platform boundary choices: "Opening-up" while still coordinating and orchestrating. In J. Furman, A. Gawer, B. Silverman, & S. Stern (Eds.), *Entrepreneurship, innovation, and platforms, advances in strategic management* (Vol. 37, pp. 227–298). Emerald Publishing.

Bouncken, R. B., Kraus, S., & Roig-Tierno, N. (2021). Knowledge-and innovation-based business models for future growth: Digitalized business models and portfolio considerations. *Review of Managerial Science, 15*(1), 1–14.
Cenamor, J., Parida, V., & Wincent, J. (2019). How entrepreneurial SMEs compete through digital platforms: The roles of digital platform capability, network capability and ambidexterity. *Journal of Business Research, 100*, 196–206.
Cenamor, J., Sjödina, D. R., & Parida, V. (2017). Adopting a platform approach in servitization: Leveraging the value of digitalization. *International Journal of Production Economics, 192*, 54–65.
Ciacci, A., Ivaldi, E., & González-Relaño, R. (2021a). A partially non-compensatory method to measure the smart and sustainable level of Italian municipalities. *Sustainability, 13*(1), 435.
Ciacci, A., Ivaldi, E., & Soliani, R. (2021b). A potential business environment of smart cities: A subjective approach. In H. Dinçer & S. Yüksel (Eds.), *Strategic outlook in business and finance innovation: Multidimensional policies for emerging economies* (pp. 11–23). Emerald Publishing Limited.
Ciampi, F., Demi, S., Magrini, A., Marzi, G., & Papa, A. (2021). Exploring the impact of big data analytics capabilities on business model innovation: The mediating role of entrepreneurial orientation. *Journal of Business Research, 123*, 1–13.
Clarysse, B., Wright, M., Bruneel, J., & Mahajan, A. (2014). Creating value in ecosystems: Crossing the chasm between knowledge and business ecosystems. *Research Policy, 43*(7), 1164–1176.
Cusumano, M. A., Gawer, A., & Yoffie, D. B. (2019). *The business of platforms: Strategy in the age of digital competition, innovation, and power*. Harper Business.
Davenport, T., Delong, D., & Beers, M. (1998). Successful knowledge management projects. *Sloan Management Review, 39*, 43–57.
De Massis, A., Frattini, F., Kotlar, J., Petruzzelli, A. M., & Wright, M. (2016). Innovation through tradition: Lessons from innovative family businesses and directions for future research. *Academy of Management Perspectives, 30*(1), 93–116.
De Reuver, M., Sørensen, C., & Basole, R. C. (2018). The digital platform: A research agenda. *Journal of Information Technology, 33*(2), 124–135.
Dominguez Gonzalez, R. V., & Massaroli de Melo, T. (2018). The effects of organization context on knowledge exploration and exploitation. *Journal of Business Research, 90*, 215–225.
Drucker, P. F. (1993). *Post capitalist society*. HarperCollins.
Eggers, F. (2020). Masters of disasters? Challenges and opportunities for SMEs in times of crisis. *Journal of Business Research, 116*, 199–208.
Eisenhardt, K. M., & Sull, D. N. (2001). Strategy as simple rules. *Harvard Business Review, 79*(1), 106–116.
Elliot, S. (2011). Transdisciplinary perspectives on environmental sustainability: A resource base and framework for IT-enabled business transformation. *MIS Quarterly, 35*, 197–236.
Eloranta, V., & Turunen, T. (2016). Platforms in service-driven manufacturing: Leveraging complexity by connecting, sharing, and integrating. *Industrial Marketing Management, 55*, 178–186.
Emerson, K., Nabatchi, T., & Balogh, S. (2012). An integrative framework for collaborative governance. *Journal of Public Administration Research and Theory, 22*(1), 1–29.
European Commission. (2019). *Annual report on European SMEs 2018/2019 Research and Development and innovation by SMEs*.
Evans, D. S., & Schmalensee, R. (2016). *Matchmakers: The new economics of multi-sided platforms*. Harvard Business Review Press.
Ferraris, A., Mazzoleni, A., Devalle, A., & Couturier, J. (2019). Big data analytics capabilities and knowledge management: Impact on firm performance. *Management Decision, 57*(8), 1923–1936.
Florida, R. (2002). The economic geography of talent. *Annals of the Association of American Geographers, 92*(4), 743–755.

Florida, R., Adler, P., King, K., & Mellander, C. (2020). The city as startup machine: The urban underpinnings of modern entrepreneurship. In *Urban studies and entrepreneurship* (pp. 19–30). Springer.

Florida, R., Gulden, T., & Mellander, C. (2008). The rise of the mega-region. *Cambridge Journal of Regions, Economy and Society, 1*(3), 459–476.

Garau, C., & Pavan, V. (2018). Evaluating urban quality: Indicators and assessment tools for smart sustainable cities. *Sustainability, 10*(3), 575.

Gawer, A. (2009). Platform dynamics and strategies: From products to services. In A. Gawer (Ed.), *Platforms, markets and innovation, chapter 3, Edward Elgar publishing*.

Gawer, A. (2021). Digital platforms' boundaries: The interplay of firm scope, platform sides, and digital interfaces. *Long Range Planning, 54*(5), 102045.

Gawer, A., & Cusumano, M. A. (2014). Industry platforms and ecosystem innovation. *Journal of Product Innovation Management, 31*(3), 417–433.

Ghazawneh, A., & Henfridsson, O. (2012). Balancing platform control and external contribution in third-party development: The boundary resources model. *Information Systems Journal, 23*(2), 173–192.

Glaeser, E. L., & Gottlieb, J. D. (2006). Urban resurgence and the consumer city. *Urban Studies, 43*(8), 1275–1299.

Glaeser, E. L., Scheinkman, J. A., & Shleifer, A. (1995). Economic growth in a cross-section of cities. *Journal of Monetary Economics, 36*(1), 117–143.

Grover, V., Chiang, R. H., Liang, T. P., & Zhang, D. (2018). Creating strategic business value from big data analytics: A research framework. *Journal of Management Information Systems, 35*(2), 388–423.

Hara, M., Nagao, T., Hannoe, S., & Nakamura, J. (2016). New key performance indicators for a smart sustainable city. *Sustainability, 8*(3), 206.

Heider, A., Gerken, M., van Dinther, N., & Hülsbeck, M. (2021). Business model innovation through dynamic capabilities in small and medium enterprises–evidence from the German Mittelstand. *Journal of Business Research, 130*, 635–645.

Helfat, C. E., & Campo-Rembado, M. A. (2016). Integrative capabilities, vertical integration, and innovation over successive technology lifecycles. *Organization Science, 27*, 249–264.

Helfat, C. E., & Raubitschek, R. S. (2018). Dynamic and integrative capabilities for profiting from innovation in digital platform-based ecosystems. *Research Policy, 47*(8), 1391–1399.

Höjer, M., & Wangel, J. (2015). Smart sustainable cities: Definition and challenges. In *ICT innovations for sustainability* (pp. 333–349). Springer.

Huang, R., Riddle, M., Graziano, D., Warren, J., Das, S., Nimbalkar, S., et al. (2016). Energy and emissions saving potential of additive manufacturing: The case of lightweight aircraft components. *Journal of Cleaner Production, 135*, 1559–1570.

Hughes, M., Filser, M., Harms, R., Kraus, S., Chang, M. L., & Cheng, C. F. (2018). Family firm configurations for high performance: The role of entrepreneurship and ambidexterity. *British Journal of Management, 29*(4), 595–612.

Huovila, A., Bosch, P., & Airaksinen, M. (2019). Comparative analysis of standardized indicators for smart sustainable cities: What indicators and standards to use and when? *Cities, 89*, 141–153.

Karimi, J., & Walter, Z. (2015). The role of dynamic capabilities in responding to digital disruption: A factor-based study of the newspaper industry. *Journal of Management Information Systems, 32*(1), 39–81.

Khan, Z., Anjum, A., Soomro, K., & Tahir, M. A. (2015). Towards cloud based big data analytics for smart future cities. *Journal of Cloud Computing Advances, Systems and Applications, 4*(2).

Kochan, T. A., & Barley, S. R. (1999). *The changing nature of work and its implications for occupational analysis*. National Research Council.

Kohtamäki, M., Parida, V., Oghazi, P., Gebauer, H., & Baines, T. (2019). Digital servitization business models in ecosystems: A theory of the firm. *Journal of Business Research, 104*, 380–392.

Ivaldi, E., Penco, L., Isola, G., & Musso, E. (2020). Smart sustainable cities and the urban knowledge-based economy: A NUTS3 level analysis. *Social Indicators Research, 150*(2), 45–72.
Jaffe, A. B., Trajtenberg, M., & Henderson, R. (1993). Geographic localization of knowledge spillovers as evidenced by patent citations. *The Quarterly Journal of Economics, 108*(3), 577–598.
Jean, R. J. B., & Kim, D. (2020). Internet and SMEs' internationalization: The role of platform and website. *Journal of International Management, 26*(1), 100690.
Jovanovic, M., Sjödin, D., & Parida, V. (2021). Co-evolution of platform architecture, platform services, and platform governance: Expanding the platform value of industrial digital platforms. *Technovation, 102218.*
Lin, F. J., & Lin, Y. H. (2016). The effect of network relationship on the performance of SMEs. *Journal of Business Research, 69*(5), 1780–1784.
Marion, T. J., Meyer, M. H., & Barczak, G. (2015). The influence of digital design and IT on modular product architecture. *Journal of Product Innovation Management, 32*(1), 98–110.
Marsal-Llacuna, M.-L., Colomer-Llinàs, J., & Meléndez-Frigola, J. (2015). Lessons in urban monitoring taken from sustainable and livable cities to better address the smart cities initiative. *Technological Forecasting and Social Change, 90*(B), 611–622.
McAfee, A., Brynjolfsson, E., Davenport, T. H., Patil, D. J., & Barton, D. (2012). Big data: The management revolution. *Harvard Business Review, 90*(10), 60–68.
Miller, K., McAdam, M., Spieth, P., & Brady, M. (2021). Business models big and small: Review of conceptualisations and constructs and future directions for SME business model research. *Journal of Business Research, 131*, 619–626.
Moos, M., Revington, N., Wilkin, T., & Andrey, J. (2019). The knowledge economy city: Gentrification, studentification and youthification, and their connections to universities. *Urban Studies, 56*(6), 1075–1092.
Musterd, S., & Gritsai, O. (2013). The creative knowledge city in Europe: Structural conditions and urban policy strategies for competitive cities. *European Urban and Regional Studies, 20*(3), 343–359.
Nambisan, S., Siegel, D., & Kenney, M. (2018). On open innovation, platforms, and entrepreneurship. *Strategic Entrepreneurship Journal, 12*(3), 354–368.
Nambisan, S., Wright, M., & Feldman, M. (2019). The digital transformation of innovation and entrepreneurship: Progress, challenges and key themes. *Research Policy, 48*(8), 103773.
Nonaka, I. (1994). A dynamic theory of organizational knowledge creation. *Organization Science, 5*(1), 14–37.
Nonaka, I., & Takeuchi, H. (1995). *The knowledge creating company: How Japanese companies create the dynamics of innovation.* Oxford University Press.
OECD. (1996). *The knowledge-based economy, in OECD*. OECD.
OECD. (2019). *An introduction to online platforms and their roles in the digital transformation.* OECD Publishing.
Oltra, V., & Saint Jean, M. (2009). Sectoral systems of environmental innovation: An application to the French automotive industry. *Technological Forecasting and Social Change, 76*(4), 567–583.
Opresnik, D., & Taisch, M. (2015). The value of big data in servitization. *International Journal of Production Economics, 165*, 174–184.
Papa, A., Dezi, L., Gregori, G. L., Mueller, J., & Miglietta, N. (2018). Improving innovation performance through knowledge acquisition: The moderating role of employee retention and human resource management practices. *Journal of Knowledge Management., 24*(3), 589–605.
Penco, L. (2015). The development of the successful city in the knowledge economy: Toward the dual role of consumer hub and knowledge hub. *Journal of the Knowledge Economy, 6*(4), 818–837.
Penco, L., Ivaldi, E., Bruzzi, C., & Musso, E. (2020). Knowledge-based urban environments and entrepreneurship: Inside EU cities. *Cities, 96*, 102443.

Penco, L., Ivaldi, E., & Ciacci, A. (2021). Entrepreneurial ecosystem and Well-being in European smart cities: A comparative perspective. *The TQM Journal, 33*(7), 318–350.
Peri, G. (2005). Determinants of knowledge flows and their effect on innovation. *Review of Economics and Statistics, 87*(2), 308–322.
Pinna, F., Masala, F., & Garau, C. (2017). Urban policies and mobility trends in Italian smart cities. *Sustainability, 9*(4), 494.
Piro, G., Cianci, I., Grieco, L. A., Boggia, G., & Camarda, P. (2014). Information centric services in smart cities. *The Journal of Systems and Software, 88*, 169–188.
Polanyi, M. (1966). *The tacit dimension*. Doubleday.
Porter, M. E. (1998). *Clusters and the new economics of competition* (Vol. 76(6), pp. 77–90). Harvard Business Review.
Powell, W. W., & Snellman, K. (2004). The knowledge economy. *Annual Review of Sociology, 30*, 199–220.
Qian, H., & Acs, Z. J. (2013). An absorptive capacity theory of knowledge spillover entrepreneurship. *Small Business Economics, 40*(2), 185–197.
Randhawa, K., Wilden, R., & Gudergan, S. (2021). How to innovate toward an ambidextrous business model? The role of dynamic capabilities and market orientation. *Journal of Business Research, 130*, 618–634.
Reis, C., Ruivo, P., Oliveira, T., & Faroleiro, P. (2020). Assessing the drivers of machine learning business value. *Journal of Business Research, 117*, 232–243.
Robertson, D., & Ulrich, K. (1998). Planning for product platforms. *Sloan Management Review, 39*(4), 19.
Romer, P. M. (1990). Human capital and growth: Theory and evidence. *Carnegie-Rochester Conference Series on Public Policy, 32*, 251.
Santoro, G., Bresciani, S., & Papa, A. (2020). Collaborative modes with cultural and creative industries and innovation performance: The moderating role of heterogeneous sources of knowledge and absorptive capacity. *Technovation, 92*, 102040.
Sawhney, M., Verona, G., & Prandelli, E. (2005). Collaborating to create: The internet as a platform for customer engagement in product innovation. *Journal of Interactive Marketing, 19*(4), 4–17.
Schumpeter, J. A. (1934). *The theory of economic development*. Oxford University Press.
Shin, D. (2009). Ubiquitous city: Urban technologies, urban infrastructure and urban informatics. *Journal of Information Science, 35*(5), 515–526.
Shree, D., Singh, R. K., Paul, J., Hao, A., & Xu, S. (2021). Digital platforms for business-to-business markets: A systematic review and future research agenda. *Journal of Business Research, 137*, 354–365.
Shrivastava, P. (1995). Environmental technologies and competitive advantage. *Strategic Management Journal, 16*(S1), 183–200.
Sirén, C., & Kohtamäki, M. (2016). Stretching strategic learning to the limit: The interaction between strategic planning and learning. *Journal of Business Research, 69*(2), 653–663.
Sirén, C., Kohtamäki, M., & Kuckertz, A. (2012). Exploration and exploitation strategies, profit performance and the mediating role of strategic learning: Escaping the exploitation trap. *Strategic Entrepreneurship Journal, 6*(1), 18–41.
Soluk, J., Kammerlander, N., & De Massis, A. (2021). *Exogenous shocks and the adaptive capacity of family firms: Exploring behavioral changes and digital technologies in the COVID-19 pandemic* (Vol. 51, p. 364). R&D Management.
Srinivasan, A., & Venkatraman, N. (2018). Entrepreneurship in digital platforms: A network-centric view. *Strategic Entrepreneurship Journal, 12*(1), 54–71.
Täuscher, K., & Laudien, S. M. (2018). Understanding platform business models: A mixed methods study of marketplaces. *European Management Journal, 36*(3), 319–329.
Tavassoli, S., Obschonka, M., & Audretsch, D. B. (2021). Entrepreneurship in cities. *Research Policy, 50*(7), 104255.
Teece, D. J. (2017). Dynamic capabilities and (digital) platform lifecycles. In *Entrepreneurship, innovation, and platforms*. Emerald Publishing Limited.

Teece, D. J. (2018). Profiting from innovation in the digital economy: Enabling technologies, standards, and licensing models in the wireless world. *Research Policy, 47*(8), 1367–1387.

Thite, M. (2011). Smart cities: Implications of urban planning for human resource development. *Human Resource Development International, 14*(5), 623–631.

Thomas, L. D. W., Autio, E., & Gann, D. M. (2014). Architectural leverage: Putting platforms in context. *Academy of Management Perspectives, 28*(2), 198–219.

Tønnessen, Ø., Dhir, A., & Flåten, B. T. (2021). Digital knowledge sharing and creative performance: Work from home during the COVID-19 pandemic. *Technological Forecasting and Social Change, 170*, 120866.

Turok, I. (2008). A new policy for Britain's cities: Choices, challenges, contradictions. *Local Economy, 23*(2), 149–166.

Van Alstyne, M., Parker, G., & Choudary, S. P. (2016). Pipelines, platforms, and the new rules of strategy. *Harvard Business Review, 94*(4), 54–62.

Yoo, Y., Boland, R. J., Jr., Lyytinen, K., & Majchrzak, A. (2012). Organizing for innovation in the digitized world. *Organization Science, 23*(5), 1398–1408.

Yoo, Y., Henfridsson, O., & Lyytinen, K. (2010). The new organizing logic of digital innovation: An agenda for information systems research. *Information Systems Research, 21*(4), 724–735.

Zackrisson, M., Avellán, L., & Orlenius, J. (2010). Life cycle assessment of lithium-ion batteries for plug-in hybrid electric vehicles–critical issues. *Journal of Cleaner Production, 18*(15), 1519–1529.

Zheng, C., Yuan, J., Zhu, L., Zhang, Y., & Shao, Q. (2020). From digital to sustainable: A scientometric review of smart city literature between 1990 and 2019. *Journal of Cleaner Production, 258*, 120689.

Chapter 4
Smart Sustainable Cities and the Urban Knowledge-Based Economy: A Practical Guidance to Monitor European Cities

Enrico Ivaldi

Abstract This chapter provides an example of using index construction techniques to measure the smart sustainable city (SSC) as a multidimensional object. SSC is a complex entity composed of multiple dimensions. SSC works correctly when its composite systems work in a symbiotic way. For this reason, it is relevant to constantly monitor the performance of the different dimensions of a SSC. Monitoring is a working method to identify strengths, weaknesses, risks, and problems, prevent cascade effects, formulate development strategies, and, more in general, govern a city. This chapter develops monitoring of SSC multiple dimensions by employing four index construction methods (e.g., Peña distance, Mazziotta and Pareto, the sum of standardized indicators, and average height). The results reveal that rankings from the different indexes are similar. Constructing monitoring through these methods is not distortive since various indexes confirm the same results. From a methodological viewpoint, the choice of the proper index should depend on the conceptual-theoretical consistency with the framework to be analyzed. The chapter ends with an analysis of the strongest and weakest SSCs.

Keywords Peña distance · Mazziotta-Pareto Index · Partially ordered set · POSET · Aggregative vs non-aggregative indexes

4.1 Introduction

The monitoring is relevant to drive SSC development. The construction of indexes is a validated approach to performing monitoring (Ciacci et al., 2021a). An index is a tool that transfers information. Its characteristics make it relatively easy to calculate, read, and interpret (Maggino, 2017a, 2017b). In these terms, the indexes can represent the primary source of information to make decisions. For instance, they help policymakers unveil hidden opportunities and extrapolate their unexpressed potential. Indexes construction embraces risk mitigation since it improves the internal capacity of a system by timely identifying risks and implementing corrective countermeasures. Indexes support the security of multisided systems by preventing cascade effects. It is particularly true for SSCs, i.e., a bundle of complex and

E. Ivaldi, A. Ciacci, *Smart Sustainable Cities and Knowledge-Based Economy*,
Human Well-Being Research and Policy Making,
https://doi.org/10.1007/978-3-031-25038-5_4

interwoven dimensions. From a long-term viewpoint, indexes assume a programmatic meaning. They assist public administration and political decision-makers in monitoring systems and supervise that the strategies aimed at developing cities generate optimal results (Marsal-Llacuna et al., 2015; Tanguay et al., 2010).

This work provides insights concerning the index construction to monitor various parameters of SSCs. In addition, it presents guidelines and suggestions on how to approach index construction. Therefore, the chapter illustrates some index construction methods to measure the attributes of SSCs. The author employs four indexes (e.g., Peña distance, Mazziotta and Pareto index, the sum of standardized indicators, and average height) (Fattore et al., 2019; Mazziotta & Pareto, 2017; Norman, 2010; Somarriba & Peña, 2009). They differ in characteristics and computational processes. The author also provides information about their features and tries to shed light on the process of index method choice. Once the author performed calculations, they discussed the results by comparing the different city-related coefficients. Overall, the results for a specific index reflect those obtained for the other indexes. In other words, a high correlation emerges between the different coefficient distributions resulting from different methods. The city rankings obtained through the application of various methods are similar. The findings tell us that the methodological choice to construct indexes should ensure "a disinterested theoretical–methodological consistency" and respect for the semantic content of the research and conceptual framework (Bonatti et al., 2021).

4.2 Index Development Guidelines

4.2.1 Peña Distance Method (DP2)

The Peña distance method (DP2) was developed by Pena (1977) starting from the Ivanovic distance (1974). In the DP2 construction, the coefficient of determination replaces the correlation coefficient. This modification changed the weighting system of the partial indicators. The coefficient of determination represents a correction factor to avoid the duplication of information deriving from the aggregation of the indicators into composite indicators or indexes.

Specifically, the DP2 is a parametric index (Somarriba & Peña, 2009). It means that its development is based on the application of a linear model of regression. In addition, DP2 construction consists in an iterative procedure that assigns weights to the partial indicators based on their correlation with a common factor, i.e., the index. The most relevant advantages of DP2 are the inter-spatial and inter-temporal comparability, the aggregation of indicators measured through diverging measurement units, the subjective weighting based on the researcher's choice, and the prevention of information duplication (Montero et al., 2010; Pena, 1977). Somarriba and Peña (2009) showed that DP2 is more robust than the traditional methods known as principal component analysis and data envelopment analysis. Bruzzi et al. (2020),

Ciacci et al., (2021a), Penco et al. (2020), and Somarriba and Peña (2009) provide practical applications of DP2 to an empirical research framework.

DP_2 is calculated as

$$I = F\ (x_1, x_2, \ldots, x_n) \tag{4.1}$$

where I is the index and n is the number of partial indicators, x_i.

The construction starts from a matrix $\underline{X}$ of order (i, j), where i (rows) indicates the number of statistical units (e.g., cities) and j (columns) indicates the number of partial indicators. The DP2 indicator expresses the distance of each statistical unit from a reference base, that is, a theoretical unit characterized by the lowest values of the studied indicators. It is defined as follows:

$$DP2_i = \sum_{j=i}^{m} \left[\left(\frac{d_{ij}}{\sigma_j} \right) \left(1 - R^2_{j,j-1,\ldots,1} \right) \right]; i = 1, 2, \ldots, n \tag{4.2}$$

where:

- i=1, 2,. . ., n are cases (statistical units).
- m is the number of indicators, X, such that $x_{ij} \in X$; $i = 1, 2, \ldots, n$; $j = 1, 2, \ldots, m$; $d = |x_{ij} - x_{\rho j}|$; $i = 1, 2, \ldots, n$; $j = 1, 2, \ldots, m$.
- ρ is the lowest value reference case—$min_i(x_{ij})$.
- σ_j indicates the standard deviation of the indicator j.
- $R^2_{j,j-1,\ldots,1}$, with $j > 1$, represents the coefficient of determination in the regression of x_j over $x_{j-1}, x_{j-2}, \ldots, x_1$.

Higher scores of DP_2 correspond to better performers.

DP_2 index is calculated through a two-step procedure:

1. We apply DP_2 for each dimension of the statistical units.
2. We apply DP_2 method another time to aggregate the different composite indicators and obtain the overall index.

Furthermore, as underlined by Somarriba and Peña (2009), Zarzosa and Sommariba (2013), Nayak and Mishra (2012), Bruzzi et al. (2020), and Penco et al. (2020), among others, the DP2 distance synthetic indicator also has the following mathematical properties:

- Existence and determination: given the mathematical function defined by DP2, it exists and takes a certain value provided that the variance of each and every one of the variables is finite and greater than zero.
- Monotony: the DP2 reacts positively to a positive variation in any of the variables and negatively to a negative transformation.
- Uniqueness quantification: for a given situation, the synthetic indicator must provide a single value or verify the invariance to changes of origin and/or scale.

Therefore, when a change is made in the scale of measurement of one or more components, the result of DP2 is not altered.

- Invariance: the DP2 is invariant to changes at origin and/or scale in the measurement of the components.
- Homogeneity: the DP2 is invariant to changes at origin and/or scale in the measurement of the components.
- Transitivity: admitting that there are three values of the synthetic indicator, if the first is greater than the second, and the second, in turn, is greater than the third, it must be verified that the first is greater than the third. This propriety is verified since DP2 is a numerical value.
- Completeness: DP2 maximizes the useful information provided by each of the simple indicators incorporated into the overall index.
- Neutrality: the weight of each single variable would be given by the useful information contained in each one. In general, it is demonstrated that the ordering of the variables in the DP2 corresponds to their relative importance, measured in terms of linear correlation with the final synthetic indicator.

4.2.2 Additive Method (Sum of Standardized Indicators (SS))

The additive method implies the sum of all the values for each indicator. When the indicators have different units of measurement, we must standardize the indicators before summing. In other words, the z-scores are calculated correspondingly to each initial indicator. Therefore, a matrix of z-scores replaces the initial indicator values during the following steps. We obtain the z-scores by subtracting the vertical average value for each statistical unit and dividing the result by the vertical average square deviation. The obtained index is an unweighted sum of z-scores (Carstairs & Morris, 1991; Forrest & Gordon, 1993; Ivaldi & Testi, 2011; Townsend, 1987).

The formula to calculate the index through the additive method consists in non-weighted sum of Z_i:

$$Z_1 = \frac{x_1 - \mu\, x_1}{\sigma\, x_1} \ldots . Z_i = \frac{x_i - \mu\, x_i}{\sigma\, x_i} \ldots Z_n = \frac{x_n - \mu\, x_n}{\sigma\, x_n} \tag{4.3}$$

where $\mu\, x_i$ and $\sigma\, x_i$ are, respectively, the means and the average square deviations of the indicators.

The additive methods can be adjusted (i.e., “stacked”) to enable a longitudinal approach making the index comparable over time (Norman, 2010; Landi et al., 2018):

$$\mu_j = \frac{\sum_{t=1}^{k} \sum_{i=1}^{n} x_{i,j,t}}{kn} \tag{4.4}$$

$$\sigma_j = \sqrt{\frac{\sum_{t=1}^{k} \sum_{i=1}^{n} \left(x_{i,j,t} - \mu_j\right)}{kn}} \tag{4.5}$$

where $i = 1, 2, \ldots, n$ are the number of statistical units, $j = 1, \ldots, m$ are the indicators, and $t = 1, \ldots, k$ are the years of the data.

4.2.3 Mazziotta and Pareto Index (MPI)

MPI is a non-compensatory aggregative method of index construction (Mazziotta & Pareto, 2017). It allows us to correct the aggregation function, i.e., the arithmetic means of the standardized indicators, by assigning a penalty to the statistical units that most diverge from the average. In other words, it assigns a penalty value to all the statistical units that have unbalanced values of their indicators (Mazziotta & Pareto, 2017). The higher the unbalance, the higher the penalty assigned. MPI is more partially non-compensatory in nature than the other existing aggregative methods. For instance, MPI generates a compensatory effect lower than the aggregative methods of arithmetic and geometric mean (Alaimo et al., 2021b). MPI implies assigning equal weights to each indicator. MPI provides a synthetic measure of a multidimensional phenomenon embracing the hypothesis that each indicator or dimension is not replaceable (Ciacci et al., 2021b).

A step-by-step process supports the MPI construction. First, it occurs the standardization and re-scaling of the basic indicators in a range (70; 130):

$$Z_{ij} = 100 \pm \frac{\left(x_{ij} - M_{xj}\right)}{S_{xj}} 10 \tag{4.6}$$

In the previous formula, M_{xj} indicates the mean of the vertical distribution, S_{xj} the standard deviation, X_{ij} the value of the jth indicator in the ith unit, and $\pm$ the sign of the relationship between the jth indicator and the phenomenon to be measured.

Second, the aggregation phase takes place. The MPI is calculated as

$$MPI_i^{+/-} = M_{zi} \pm S_{zi} CV_i \tag{4.7}$$

where M_{zi}, S_{zi}, and CV_i are, respectively, the horizontal mean (i.e., the mean considering all the standardized values corresponding to a specific statistical unit), standard deviation, and coefficient of variation of the unit i (Mazziotta & Pareto, 2017).

As a result, we obtain a value corresponding to the multiple dimensions that represent a phenomenon for all the statistical units. We can obtain the final MPI value for each unit by aggregating the multiple dimensions.

MPI is a proper element to introduce the debate between aggregative and non-aggregative methods. In the literature, the debate on the adaptability of aggregative methods compared to non-aggregative methods to analyze multidimensional systems emerged. Some works explored the weaknesses of the aggregative methods (Maggino, 2017a; Fattore, 2017a; Freudenberg, 2003). These methods raise conceptual and methodological issues. Non-aggregative methods have gained prominence since they overcome some limits characterizing aggregative methods (Alaimo et al., 2021a, Alaimo et al., 2021b; Ivaldi et al., 2020a; Fattore, 2018; Maggino, 2017b). MPI stands in the middle ground between aggregative and non-aggregative methods. MPI represents a solution to partially overcome the compensability issues related to the aggregative methods (Mazziotta & Pareto, 2020). The penalty effect counterbalances potential distortive effects produced by the compensation during the aggregation phase. More in general, MPI has the following characteristics:

- It assigns the same weight to all the partial indicators.
- It is applicable to phenomenon measured both positively (higher values of the indicators correspond to better performance) and negatively (higher values of the indicators correspond to worse performance) by varying the penalty sign.
- It can be decomposed into its constituent parts (the average effect, the compensatory part, and the penalty effect) to explore the measurement composition and the statistical units more in-depth.
- It enables "relative" (not absolute) comparisons between statistical units.

In addition, the aggregative methods represent consolidated approaches to studying multidimensional phenomena despite their limitations. Many empirical examples exist concerning the application of the aggregative approach to study multidimensional systems (Ciacci et al., 2021b, 2021c; Ivaldi et al., 2020b, 2020c; Mazziotta & Pareto, 2020; Penco et al., 2020). For this reason, aggregative methods ensure robustness and recognizability among scholars.

4.2.4 Partially Ordered Set (POSET)

POSET is a non-aggregative approach commonly used to study inequality (Arcagni et al., 2018; Ivaldi et al., 2020a; Alaimo et al., 2022a, 2022b). However, its design allows us to analyze other phenomena through comparison. POSET stays for "partially ordered set," i.e., a set of tools designed to describe and treat order relations. POSET assumes that the aggregative methods may alter or partially describe complex and multidimensional phenomena. In this regard, the issue refers to the notion of compensation, i.e., the balancing of the high values of some indicators with the low values of other indicators. The arithmetic mean is an example of a compensative method. The proper characteristics of the different dimensions limit their substitutability. Therefore, it is impossible to proceed with compensation without producing distortion in the analysis (Sen, 1992). The POSET method offers a solution to compensation-related problems since it ensures the preservation of specificity of the cases according to their possible incomparability. In addition,

POSET allows us to process data without altering its primary nature. In fact, POSET does not imply any standardization or normalization procedure. POSET method is appropriate to analyze complex and multidimensional phenomena from a cross-sectional or longitudinal perspective (e.g., temporal POSET, Alaimo et al., 2021a). All these elements together make the evaluation process more robust.

The literature provides examples of empirical studies through POSET. Arcagni et al. (2018) analyzed the deprivation of migrants in Lombardy to identify the situations of poverty and fragility affecting the different ethnic groups. Ivaldi et al., (2020a) studied Argentinian urban deprivation from a multidimensional perspective to establish the political intervention priority. Fattore and Arcagni et al. (2018) applied the POSET to study child well-being in the Democratic Republic of the Congo's regions. Wittmann and Brüggemann (2014) used the POSET method to identify the optimal environmental and technical parameters characterizing different types of cars. Hilckmann et al. (2017) studied the relationship between political parties and sustainable development in Germany. Penco et al. (2021) apply POSET to investigate the relationship between innovative entrepreneurial ecosystems and perceived well-being in European smart cities. These examples give us an idea of the application field of the POSET method. Alaimo et al. (2022b) describe each well-being domain as a partially ordered set and compute synthetic indicators for well-being rankings at regional level for year 2017.

From a mathematical point of view, a partially ordered set (or POSET) is a set π equipped with a partial order relation $\leq$, i.e., a binary relation satisfying the properties of reflexivity, antisymmetry, and transitivity (Davey & Priestley, 2002; Neggers & Kim, 1998). A profile is associated with each statistical unit of the distribution. In the POSET method, the sequence of statistical units is called ordered structure, that is, a sequence of integers within a specified range. A range is a set of integers, that is, a numerical scale containing all the values of the different indicators. The POSET admits two different situations. First, we can find that two statistical units are comparable (e.g., comparability). In this case, we can compare the statistical units to establish their dominance relationship ($b \leq a$ or $a \leq b$). In the case of comparability between statistical units, we refer to the notion of linear order or complete order. Second, we can have incomparability between statistical units. In other words, we cannot establish a precise relationship between two or more statistical units because their profiles make them incomparable. Mutually comparable elements form a subset known as chain. Mutually incomparable elements originate a subset that is called antichain ($b \parallel a$).

Given $b, a \in \pi$, a covers b ($b \prec a$) if $b \leq a$ and there is no other element $c \in \pi$ such that $b \leq c \leq a$. A statistical unit $b \in \pi$ such that $b \leq a$ implies $b = a$ is called maximal; if for each $a \in \pi$ it is $a \leq b$, then b is called maximum or the greatest statistical unit of π. A statistical unit $b \in \pi$ such that $a \leq b$ implies $b = a$ is instead called minimal; if for each $a \in \pi$ it is $b \leq a$, then b is called (the) minimum or the least element of π. Given $b \in \pi$, the down-set of b (written $\downarrow b$) is the set of all the statistical units $a \in \pi$ such that $a \leq b$. The up-set of b (written $\uparrow b$) is the set of all the statistical units $a \in \pi$ such that $b \leq a$. Considering two partially ordered sets π and τ on the same set, τ is an extension of π if $b \leq_\pi a$ in π implies $b \leq_\tau a$ in τ. Said differently, τ is an extension of π if it may be obtained from the latter turning some incomparabilities into comparabilities. An extension of a complete order is called a linear extension. The set of linear extensions of a POSET π is denoted by $\Omega(\pi)$ (Fattore, 2017b).

A set is defined as linearly (or totally) ordered when all its paired elements are linked by an order relation. The Hasse diagram is the typical way of graphical representation to describe a partial relationship. It is an acyclic oriented graph, drawn due to the following rules:

- If $b \leq a$, the node b is placed lower than the node a.
- If $b \prec a$, an edge connects them.
- The edge always reads from top to bottom.

Therefore, nodes each other connected by a downward path form the Hasse diagram. Unconnected nodes indicate incomparability between statistical units (Arcagni et al., 2018; Carlsen & Bruggemann, 2017).

POSET provides a complete order out of a partial order (Fattore, 2017b). The procedure implies that after having assigned the evaluation scores to all the statistical units of a sequence, the statistical units can be linearly ordered. The average height assigns a score in the range [0, 1] to each unit of a sequence. The score describes the statistical unit's position along a low-high axis (Fattore et al., 2019). The average height computation is a multistep procedure. The first step consists in listing all the linear extensions of the sequence. For each linear extension, the height of the statistical units is calculated as 1 plus the number of elements below x in the linear order. Finally, we must calculate the average height of each statistical unit on its linear extensions, that is, the arithmetic mean of the heights of x in all linear extensions (Winkler, 1982).

Specifically, the average height procedure of calculation is the following (Fattore & Arcagni, 2018):

- Extracting all the linear extensions of a certain order structure π to create $\Omega(\pi)$.
- Assigning the rank $r_\ell(b)$ of b in ℓ, defined as 1 + the number of edges linking b to the maximum of ℓ, to each statistical unit $b \in \pi$ and $\ell \in \Omega(\pi$.)
- Calculating the average $r(b)$ of $r_\ell(b)$ over $\Omega(\pi)$ of each $b \in \pi$
- Obtaining the average rank of each statistical unit belonging to the distribution
- Subtracting n + 1, where n is the number of statistical units, to the average rank value of each statistical unit to obtain the average height value of each statistical unit

4.3 Data Description

Table 4.1 shows the European cities of the sample. It involves all the cities sampled by the Eurostat survey. The exclusion of some cities occurred because of missing data. Overall, 85 cities form the sample.

Table 4.2 provides a description of the variables used. They are all in percentage form and come from the Eurostat database. All the variables refer to the city level

Table 4.1 Country description

Country	City (Eurostat codes)	City (labels)
Belgium	BE001C1	Bruxelles/Brussels
Belgium	BE002C1	Antwerpen
Belgium	BE005C1	Liège
Bulgaria	BG001C1	Sofia
Bulgaria	BG004C1	Burgas
Czech Republic	CZ001C1	Praha
Czech Republic	CZ003C1	Ostrava
Denmark	DK001K2	København (greater city)
Denmark	DK004C2	Aalborg
Germany	DE001C1	Berlin
Germany	DE002C1	Hamburg
Germany	DE003C1	München
Germany	DE004C1	Köln
Germany	DE005C1	Frankfurt am Main
Germany	DE006C1	Essen
Germany	DE007C1	Stuttgart
Germany	DE008C1	Leipzig
Germany	DE009C1	Dresden
Germany	DE010C1	Dortmund
Germany	DE011C1	Düsseldorf
Estonia	EE001C1	Tallinn
Estonia	EE002C1	Tartu
Estonia	EE003C1	Narva
Ireland	IE001K1	Dublin (greater city)
Greece	EL001K1	Athina (greater city)
Greece	EL004C1	Irakleio
Spain	ES001C1	Madrid
Spain	ES002K2	Barcelona
Spain	ES006C1	Málaga
Spain	ES013C1	Oviedo
France	FR001K1	Paris (greater city)
France	FR006C2	Strasbourg
France	FR007C1	Bordeaux
France	FR009C1	Lille
France	FR013C2	Rennes
France	FR203C1	Marseille
Croatia	HR001C1	Zagreb
Italy	IT001C1	Roma
Italy	IT003K1	Napoli (greater city)
Italy	IT004C1	Torino
Italy	IT005C1	Palermo
Italy	IT009C1	Bologna

(continued)

Table 4.1 (continued)

Country	City (Eurostat codes)	City (labels)
Italy	IT012C1	Verona
Cyprus	CY001C1	Lefkosia
Latvia	LV001C1	Riga
Lithuania	LT001C1	Vilnius
Luxembourg	LU001C1	Luxembourg
Hungary	HU001C1	Budapest
Hungary	HU002C1	Miskolc
Malta	MT001C1	Valletta
The Netherlands	NL002C2	Greater Amsterdam
The Netherlands	NL003C2	Greater Rotterdam
The Netherlands	NL007C1	Groningen
Austria	AT001C1	Wien
Austria	AT002C1	Graz
Poland	PL001C1	Warszawa
Poland	PL003C1	Kraków
Poland	PL006C1	Gdansk
Poland	PL011C1	Bialystok
Portugal	PT001K1	Lisboa (greater city)
Portugal	PT003C1	Braga
Romania	RO001C1	Bucuresti
Romania	RO002C1	Cluj-Napoca
Romania	RO011C1	Piatra Neamt
Slovenia	SI001C1	Ljubljana
Slovakia	SK001C1	Bratislava
Slovakia	SK002C1	Kosice
Finland	FI001K2	Helsinki/Helsingfors (greater city)
Finland	FI004C3	Oulu
Sweden	SE001K1	Stockholm (greater city)
Sweden	SE003C1	Malmö
Iceland	IS001C1	Reykjavík
Norway	NO001C1	Oslo
Switzerland	CH001K1	Zürich (greater city)
Switzerland	CH002K1	Genève (greater city)
United Kingdom	UK001K2	London (greater city)
United Kingdom	UK004C1	Glasgow City
United Kingdom	UK008K1	Greater Manchester
United Kingdom	UK009C1	Cardiff
United Kingdom	UK012C2	Belfast
United Kingdom	UK013K1	Tyneside conurbation
Turkey	TR001C1	Ankara
Turkey	TR003C1	Antalya
Turkey	TR007C1	Diyarbakir
Turkey	TR012C1	Istanbul

Table 4.2 Variables description

Source	Dimension	Code	Year	Level	Label	Form
Eurostat	ICT and transport	PS1012V	2019	City	Public transport in the city, for example bus, tram, or metro: very satisfied	Percentage
Eurostat	ICT and transport	PS1013V	2019	City	Public transport in the city, for example bus, tram, or metro: rather satisfied	Percentage
Eurostat	ICT and transport	PS3547V	2019	City	Public transport Affordable: strongly agree	Percentage
Eurostat	ICT and transport	PS3548V	2019	City	Public transport Affordable: somewhat agree	Percentage
Eurostat	ICT and transport	PS3552V	2019	City	Public transport Safe: strongly agree	Percentage
Eurostat	ICT and transport	PS3553V	2019	City	Public transport Safe: somewhat agree	Percentage
Eurostat	ICT and transport	PS3557V	2019	City	Public transport Easy to get: strongly agree	Percentage
Eurostat	ICT and transport	PS3558V	2019	City	Public transport Easy to get: somewhat agree	Percentage
Eurostat	ICT and transport	PS3562V	2019	City	Public transport Frequent (comes often): strongly agree	Percentage
Eurostat	ICT and transport	PS3563V	2019	City	Public transport Frequent (comes often): somewhat agree	Percentage
Eurostat	ICT and transport	PS3567V	2019	City	Public transport Reliable (comes when it says it will): strongly agree	Percentage
Eurostat	ICT and transport	PS3568V	2019	City	Public transport Reliable (comes when it says it will): somewhat agree	Percentage
Eurostat	ICT and transport	ISOC_SK_DSKL_I	2019	Country	Individuals' level of digital skills	Percentage
Eurostat	ICT and transport	ISOC_SK_CSKL_I	2019	Country	Individuals' level of computer skills	Percentage

(continued)

Table 4.2 (continued)

Source	Dimension	Code	Year	Level	Label	Form
Eurostat	ICT and transport	ISOC_CI_IN_H	2021	Country	Households—level of Internet access	Percentage
Eurostat	People	PS1022V	2015	City	Schools in the city: very satisfied	Percentage
Eurostat	People	PS1023V	2015	City	Schools in the city: rather satisfied	Percentage
Eurostat	People	PS1506V	2019	City	Healthcare services, doctors and hospitals: very satisfied	Percentage
Eurostat	People	PS1507V	2019	City	Healthcare services, doctors and hospitals: rather satisfied	Percentage
Eurostat	People	PS1062V	2019	City	Sports facilities such as sport fields and indoor sport halls in the city: very satisfied	Percentage
Eurostat	People	PS1063V	2019	City	Sports facilities such as sport fields and indoor sport halls in the city: rather satisfied	Percentage
Eurostat	People	PS1082V	2019	City	Cultural facilities such as concert halls, theaters, museums, and libraries in the city: very satisfied	Percentage
Eurostat	People	PS1083V	2019	City	Cultural facilities such as concert halls, theaters, museums, and libraries in the city: rather satisfied	Percentage
Eurostat	People	PS2092V	2015	City	You are satisfied to live in this city: strongly agree	Percentage
Eurostat	People	PS2093V	2015	City	You are satisfied to live in this city: somewhat agree	Percentage
Eurostat	People	PS2012V	2019	City	In this city it is easy to find a good job: strongly agree	Percentage
Eurostat	People	PS2013V	2019	City	In this city it is easy to find a good job: somewhat agree	Percentage
Eurostat	People	PS2022V	2015	City	Foreigners who live in this city are well integrated: strongly agree	Percentage
Eurostat	People	PS2023V	2015	City	Foreigners who live in this city are well integrated: somewhat agree	Percentage
Eurostat	People	PS2032V	2019	City	In this city, it is easy to find good housing at a reasonable price: strongly agree	Percentage
Eurostat	People	PS2033V	2019	City	In this city, it is easy to find good housing at a reasonable price: somewhat agree	Percentage
Eurostat	People	PS3062V	2019	City	Public spaces in this city such as markets, squares, pedestrian areas: very satisfied	Percentage
Eurostat	People	PS3063V	2019	City	Public spaces in this city such as markets, squares, pedestrian areas: rather satisfied	Percentage

Eurostat	People	PS3092V	2019	City	Generally speaking, most people in this city can be trusted: strongly agree	Percentage
Eurostat	People	PS3093V	2019	City	Generally speaking, most people in this city can be trusted: somewhat agree	Percentage
Eurostat	People	PS3230V	2019	City	State of streets and buildings in my neighborhood: very satisfied	Percentage
Eurostat	People	PS3231V	2019	City	State of streets and buildings in my neighborhood: rather satisfied	Percentage
Eurostat	People	PS3250V	2015	City	Availability of retail shops: very satisfied	Percentage
Eurostat	People	PS3251V	2015	City	Availability of retail shops: rather satisfied	Percentage
Eurostat	People	PS3340V	2019	City	The financial situation of your household: very satisfied	Percentage
Eurostat	People	PS3341V	2019	City	The financial situation of your household: fairly satisfied	Percentage
Eurostat	People	PS3290V	2015	City	You feel safe in this city: strongly agree	Percentage
Eurostat	People	PS3291V	2015	City	You feel safe in this city: somewhat agree	Percentage
Eurostat	Sustainability	PS1052V	2019	City	Green spaces such as public parks or gardens: very satisfied	Percentage
Eurostat	Sustainability	PS1053V	2019	City	Green spaces such as public parks or gardens: rather satisfied	Percentage
Eurostat	Sustainability	PS3122V	2019	City	This city is committed to the fight against climate change (e.g.; reducing energy consumption in housing or promoting alternatives to transport by car): strongly agree	Percentage
Eurostat	Sustainability	PS3123V	2019	City	This city is committed to the fight against climate change (e.g.; reducing energy consumption in housing or promoting alternatives to transport by car): somewhat agree	Percentage
Eurostat	Sustainability	PS3260V	2019	City	The quality of the air in the city: very satisfied	Percentage
Eurostat	Sustainability	PS3261V	2019	City	The quality of the air in the city: rather satisfied	Percentage
Eurostat	Sustainability	PS3270V	2019	City	The noise level in the city: very satisfied	Percentage
Eurostat	Sustainability	PS3271V	2019	City	The noise level in the city: rather satisfied	Percentage
Eurostat	Sustainability	PS3280V	2019	City	The cleanliness in the city: very satisfied	Percentage
Eurostat	Sustainability	PS3281V	2019	City	The cleanliness in the city: rather satisfied	Percentage
Eurostat	Government	PS2042V	2015	City	When you contact administrative services of this city, they help you efficiently: strongly agree	Percentage
Eurostat	Government	PS2043V	2015	City	When you contact administrative services of this city, they help you efficiently: somewhat agree	Percentage

(continued)

Table 4.2 (continued)

Source	Dimension	Code	Year	Level	Label	Form
Eurostat	Government	PS3320V	2015	City	The public administration of the city can be trusted: strongly agree	Percentage
Eurostat	Government	PS3321V	2015	City	The public administration of the city can be trusted: somewhat agree	Percentage
Eurostat	Government	PS3572V	2019	City	Confidence in the local police force: yes	Percentage
Eurostat	Government	PS3587V	2019	City	I am satisfied with the amount of time it takes to get a request solved by my local public administration: strongly agree	Percentage
Eurostat	Government	PS3588V	2019	City	I am satisfied with the amount of time it takes to get a request solved by my local public administration: somewhat agree	Percentage
Eurostat	Government	PS3592V	2019	City	The procedures used by my local public administration are straightforward and easy to understand: strongly agree	Percentage
Eurostat	Government	PS3593V	2019	City	The procedures used by my local public administration are straightforward and easy to understand: somewhat agree	Percentage
Eurostat	Government	PS3601V	2019	City	The fees charged by my local public administration are reasonable: strongly agree	Percentage
Eurostat	Government	PS3602V	2019	City	The fees charged by my local public administration are reasonable: somewhat agree	Percentage
Eurostat	Government	PS3606V	2019	City	Information and services of my local public administration can be easily accessed online: strongly agree	Percentage
Eurostat	Government	PS3607V	2019	City	Information and services of my local public administration can be easily accessed online: somewhat agree	Percentage
Eurostat	Government	PS3613V	2019	City	There is corruption in my local public administration: somewhat disagree	Percentage
Eurostat	Government	PS3614V	2019	City	There is corruption in my local public administration: strongly disagree	Percentage

with the exception of "Individuals' level of digital skills," "Individuals' level of computer skills," and "Households—level of Internet access" that indicate the national level data. The author replicated the national data for all the cities of the same country to overcome the lack of ICT data at the city level. The variables reflect the four higher-order dimensions described in Chap. 1, i.e., ICT and transport, People, Sustainability, and Government. The variables mainly refer to the 2019, 2015, and 2021 years.

4.4 Results and Discussions

Table 4.3 highlights a high correlation between the ranks of the indicators. In other words, the correlation values indicate many similarities between the ranks of the cities. Considering Kendall's tau index, the correlations range from a minimum value of 0.813 (DP2-AH) to a maximum of 0.932 (DP2-SS). It highlights that the results obtained through different index construction methods are similar. The finding demonstrates a relatively high degree of consistency in the results. Applying different indexes does not produce distortive effects on the results. The method of index construction adopted is not a determinant of the results by itself.

Table 4.4 reports the results of index calculations expressed as coefficients and the relative city rank. The computed indexes are P2 distance (DP2), Mazziotta and Pareto index (MPI), sum of standardized indicators (SS), and average height (AH). In the following paragraphs, we discuss the overall results obtained from the different methods more in-depth. We also discuss the results of each city dimension in Appendix (Tables 4.5, 4.6, 4.7, and 4.8).

The European Union has shown an increasing focus on smart cities. This is evidenced by the creation of a dedicated platform focused on them: the Smart Cities Marketplace Platform. This platform aims to unite different cities and stakeholders into a strong network in a way that improves the quality of life for citizens and healthily increases the competitiveness of European cities and industries (European Commission, 2022). To do this, the platform acts by following three distinct phases:

Table 4.3 Kendall and Spearman correlation coefficient between indexes

		DP2	MPI	SS	AH
Kendall's tau	DP2	1.000	0.889**	0.932**	0.813**
	MPI	0.889**	1.000	0.929**	0.826**
	SS	0.932**	0.929**	1.000	0.838**
	AH	0.813**	0.826**	0.838**	1.000
Spearman's rho	DP2	1.000	0.978**	0.991**	0.950**
	MPI	0.978**	1.000	0.990**	0.959**
	SS	0.991**	0.990**	1.000	0.964**
	AH	0.950**	0.959**	0.964**	1.000

Table 4.4 Indexes coefficients and ranks

Country	DP2	DP2 ranks	MPI	MPI ranks	SS	SS ranks	AH	AH ranks	AR
Aalborg	9.207	2	108.299	4	31.528	3	80.540	3	3.000
Ankara	5.749	64	98.515	50	−6.300	60	47.375	39	53.250
Antalya	5.439	65	96.257	66	−12.250	65	18.363	71	66.750
Antwerpen	7.657	29	102.443	32	11.769	31	56.076	31	30.750
Athina (greater city)	1.153	84	82.675	82	−56.912	82	3.702	82	82.500
Barcelona	5.804	63	97.195	58	−6.531	62	26.251	62	61.250
Belfast	8.088	18	105.431	13	19.806	13	73.670	10	13.500
Berlin	5.913	60	97.632	57	−6.182	58	25.885	63	59.500
Bialystok	8.132	14	105.055	18	17.307	22	67.499	15	17.250
Bologna	6.083	54	98.040	51	−5.509	55	30.421	52	53.000
Bordeaux	8.018	22	104.703	22	16.328	25	60.046	25	23.500
Braga	7.116	39	100.821	40	4.748	41	43.685	44	41.000
Bratislava	4.599	71	91.999	76	−23.236	75	22.720	67	72.250
Bruxelles/Brussels	6.319	50	98.676	49	−2.508	49	29.556	55	50.750
Bucuresti	3.547	78	88.999	79	−33.562	78	11.955	75	77.500
Budapest	5.301	67	95.314	68	−15.283	68	14.707	72	68.750
Burgas	6.154	52	97.066	60	−6.372	61	27.581	61	58.500
Cardiff	8.860	5	107.210	7	26.033	6	80.198	5	5.750
Cluj-Napoca	7.136	38	100.105	42	6.901	38	58.052	27	36.250
Diyarbakir	5.345	66	96.007	67	−13.138	67	29.114	57	64.250
Dortmund	6.683	45	101.032	39	3.440	43	46.031	42	42.250
Dresden	7.846	25	104.017	24	16.826	24	65.976	18	22.750
Dublin (greater city)	7.447	32	102.807	30	10.884	32	47.957	38	33.000
Düsseldorf	8.095	17	105.169	15	19.228	15	66.456	16	15.750
Essen	6.404	48	99.855	43	−0.005	45	34.959	49	46.250
Frankfurt am Main	7.239	35	101.895	38	9.230	37	49.683	35	36.250

Gdansk	7.617	30	102.392	33	10.419	33	57.576	28	31.000
Genève (greater city)	8.710	7	106.817	8	25.260	8	73.940	9	8.000
Glasgow City	8.206	12	104.718	21	18.215	18	59.331	26	19.250
Graz	8.636	9	106.387	9	24.467	9	74.002	8	8.750
Greater Amsterdam	7.982	23	102.322	34	15.593	26	47.372	40	30.750
Greater Manchester	7.912	24	104.872	19	16.988	23	64.162	22	22.000
Greater Rotterdam	8.305	10	102.863	29	17.421	21	54.791	33	23.250
Groningen	9.174	3	108.032	5	30.108	4	77.718	7	4.750
Hamburg	7.518	31	103.883	25	13.603	28	55.618	32	29.000
Helsinki/Helsingfors (greater city)	8.226	11	106.320	10	22.983	10	66.429	17	12.000
Irakleio	3.170	79	88.779	80	−35.364	79	10.521	77	78.750
Istanbul	3.021	80	89.876	78	−35.627	80	6.869	81	79.750
København (greater city)	8.163	13	105.663	12	21.532	11	63.982	23	14.750
Köln	6.062	55	96.760	65	−5.494	54	30.392	53	56.750
Kosice	6.174	51	97.859	53	−5.749	56	22.487	68	57.000
Kraków	6.788	44	97.782	55	−1.408	48	32.811	51	49.500
Lefkosia	4.578	74	93.182	73	−21.976	73	14.318	73	73.250
Leipzig	7.168	37	102.715	31	9.345	36	47.347	41	36.250
Liège	5.994	58	97.956	52	−5.845	57	25.418	64	57.750
Lille	6.386	49	99.622	45	−1.025	47	28.702	59	50.000
Lisboa (greater city)	4.353	77	92.635	75	−24.813	76	7.850	80	77.000
Ljubljana	7.205	36	102.884	28	9.479	35	48.434	36	33.750
London (greater city)	7.404	33	103.272	27	12.248	30	51.130	34	31.000
Luxembourg	9.122	4	109.377	2	32.227	2	78.166	6	3.500
Madrid	4.580	73	93.367	72	−19.051	71	24.659	65	70.250
Málaga	6.519	46	98.769	48	−0.520	46	45.915	43	45.750
Malmö	8.035	20	105.142	17	17.564	20	68.466	14	17.750

(continued)

Table 4.4 (continued)

Country	DP2	DP2 ranks	MPI	MPI ranks	SS	SS ranks	AH	AH ranks	AR
Marseille	4.537	75	93.147	74	−21.995	74	8.636	79	75.500
Miskolc	5.238	68	97.034	62	−12.582	66	27.983	60	64.000
München	8.124	16	105.749	11	20.475	12	65.913	19	14.500
Napoli (greater city)	1.599	82	80.245	84	−62.958	83	3.469	83	83.000
Narva	4.977	69	93.512	71	−18.765	70	22.954	66	69.000
Oslo	7.715	27	104.437	23	19.019	17	64.923	20	21.750
Ostrava	6.950	41	99.729	44	3.499	42	40.526	46	43.250
Oulu	7.762	26	101.973	37	12.324	29	64.918	21	28.250
Oviedo	7.305	34	102.303	35	9.514	34	56.428	30	33.250
Palermo	0.486	85	77.799	85	−72.450	85	2.354	84	84.750
Paris (greater city)	6.138	53	97.730	56	−4.709	53	19.387	70	58.000
Piatra Neamt	5.880	61	98.901	47	−4.444	52	41.575	45	51.250
Praha	7.077	40	99.587	46	5.073	40	56.975	29	38.750
Rennes	8.770	6	107.297	6	25.397	7	82.060	2	5.250
Reykjavík	5.971	59	97.064	61	−4.387	51	35.558	48	54.750
Riga	4.589	72	93.813	70	−19.821	72	11.724	76	72.500
Roma	1.349	83	80.290	83	−63.739	84	2.193	85	83.750
Sofia	2.880	81	87.708	81	−39.423	81	12.124	74	79.250
Stockholm (greater city)	8.069	19	105.165	16	19.347	14	70.249	13	15.500
Strasbourg	8.019	21	104.763	20	17.662	19	72.445	11	17.750
Stuttgart	7.665	28	103.294	26	14.358	27	48.133	37	29.500
Tallinn	6.948	42	102.031	36	6.546	39	61.070	24	35.250
Tartu	6.512	47	97.789	54	−2.891	50	37.793	47	49.500
Turin	4.361	76	91.220	77	−28.981	77	10.371	78	77.000
Tyneside conurbation	8.128	15	105.333	14	19.073	16	72.220	12	14.250

Valletta	4.878	70	93.911	69	−16.017	69	29.031	58	66.500
Verona	5.828	62	96.804	64	−8.644	64	29.385	56	61.500
Vilnius	6.807	43	100.762	41	2.503	44	33.309	50	44.500
Warszawa	5.999	57	97.104	59	−6.259	59	29.605	54	57.250
Wien	8.707	8	108.745	3	28.845	5	80.236	4	5.000
Zagreb	6.022	56	96.879	63	−7.104	63	20.649	69	62.750
Zürich (greater city)	10.235	1	112.655	1	44.714	1	83.940	1	1.000

DP2 P2 distance; *MPI* Mazziotta-Pareto index; *SS* sum of standardized indicators; *AH* average height; *AR* average rank between DP2, MPI, SS, AH ranks

- *Explore*: all smart and green projects of European cities are made available in a vast database. It enhances best practices and creates a vast fertile ground for the creation of new ideas.
- *Shape*: in this phase, the policymaker is guided step by step in the implementation of ideas into actual projects. This phase is critical to attracting investment from both public and private stakeholders.
- *Deal*: the final phase in which the idea becomes concrete and in which a one-to-one exchange with other community members takes place.

Key areas of focus for the platform include urban mobility, the creation of increasingly sustainable energy infrastructure, a strong focus on citizen well-being, accountability, and the creation of public-private partnerships (European Commission, 2022).

This growing focus on smart cities is reflected in the implementation of more and more sustainable and smart solutions by individual European cities. As seen in Table 4.4, the smartest city is Zurich (AR = 1). This ranking is likely due to the strategy implemented by the Zurich City Council: the Strategy Smart City Zurich (Zurich City Council, 2018). This measure, explicated in several urban digitization strategies, has three primary goals: meeting the future needs of the population, promoting innovation, and positioning the city as a smart city. The measure provides for close collaboration between politics, business, science, culture, and society to achieve these three ambitious goals. It will lead to the full digitization of the city, which will enable all citizens to actively participate in government through their feedback. The Strategy Smart City Zurich is also expressed in four guiding principles (Zurich City Council, 2018):

1. Focus on certain categories of citizens and citizens and the city's future challenges
2. Networking among different stakeholders both public and private
3. Accountability, availability, and GDPR compliance of data
4. Agile innovation and development

Finally, to implement this plan, the Zurich City Council has provided several tools (Zurich City Council, 2018):

- Public grants to encourage innovation and the implementation of new projects
- Bottom-up approach in the development of new ideas
- Special scholarships on innovation issues
- The creation of a laboratory on smart cities
- Both national and international collaborations
- Accountability and data transparency
- Preparation of an annual report on the city's progress

Then follow in the top 10 of the most virtuous European cities in terms of implementing smart solutions Aalborg (AR = 3), Luxembourg (AR = 3.5), Groningen (AR = 4.75), Wien (AR = 5), Rennes (AR = 5.25), Cardiff (AR = 5.75), Genève (AR = 8), Graz (AR = 8.75), and Helsinki (AR = 12).

The second-ranked city, Aalborg, has been investing in sustainable issues for about 30 years. In fact, in 1994 and 2004, the Danish city council signed the Aalborg Charter and the Aalborg Commitments, respectively. Thanks to these measures, the policy framework for sustainability issues in the city could be defined. In addition, the Smart Aalborg plan has recently been implemented, which aims to guarantee jobs for all citizens. In addition, this plan includes the creation of places for the exchange of ideas and citizen action. It has created fertile ground for the implementation of holistic solutions to digitization, innovation, and ICT issues (Smart City Press, 2020).

The third-ranked city, Luxembourg, has also been positioned as a smart city in Europe for years due to its policies in this area. In fact, the city's City Council recently focused on five focus areas (Etienne, 2019):

1. *Smart People*: an online platform in three different languages (French, English, and German) has been implemented to increase culture and sociability and make citizens aware of upcoming events.
2. *Smart Governance*: the city makes available and accessible all data pertaining to the general development plan, population status, or housing prices by neighborhood to ensure fair transparency and accountability. In addition, every session of the City Council is live streamed with simultaneous translation into German sign language.
3. *Smart Mobility*: the city provides access to a range of useful, real-time data through a free app to facilitate mobility.
4. *Smart Environment*: the city has implemented an interactive tool that lets people know what type of sustainable energy can be applied to their homes, the cost of the system, and the possibility of having government subsidies.
5. *Smart Living*: some augmented reality projects in some districts in the city are being implemented.

The city of Groningen, on the other hand, together with the city of Oulu, was chosen as the flagship city of the European Making-City project launched in 2018. This project aims to create smart cities with low carbon emissions through a transformation of the energy system. The European Union chose this city for its policies aimed at changing urban energy sources. In fact, the Netherlands has for years based its energy sources on natural gas. However, the consequences caused by climate change have shown the need to find new and sustainable energy sources. For this reason, the City Council of the City of Groningen adopted a Master Plan in 2011 with the goal of energy neutrality by 2035 (Making City, 2019).

Like Groningen, the city of Vienna is also making a strong commitment to finding solutions against climate change. In fact, in 2014 the Austrian city council launched its "Smart City Wien" strategy, which will have 2050 as its timeframe. In 2020, moreover, through a government agreement, Vienna anchored the goal of achieving climate neutrality by 2040. The mission of the "Smart City Wien" strategy is: "High quality of life for everyone in Vienna through social and technical innovation in all areas while maximizing conservation of resources." From this, we can see the three main areas of focus: quality of life, conservation of resources, and innovation.

Therefore, the Austrian city's policies are focused on social inclusion, youth policies, energy conservation, combating climate change, and making Vienna the capital of innovation and digitalization (Smart City Wien, 2022).

The strategies of the sixth-ranked city, Rennes, are based on transparency and full participation of its citizens in city life. In fact, the French city was the first local authority to make its data accessible to the population in 2010. Since then, Rennes has been committed to implementing increasingly innovative and smart solutions. There are three main projects active at the moment: (i) Rennes métropole in open data; (ii) metropolitan transportation public data service; and (iii) 3DExperienCity (Rennes Métropole, 2022).

Regarding the first project, the city of Rennes has created a web platform to disseminate its public data. Specifically, the data disseminated cover the geolocation of public facilities, city budgets, transportation, cultural and sporting events, the marital status of female citizens, and statistics on waste collection or equipment use. In addition, the city provides real-time data regarding traffic status, public transportation network, and parking availability. As for the second project currently underway, however, in February 2018, the French city launched a metropolitan data service to collect and share data of general interest so that new and innovative urban services can be created. The themes chosen to test this solution are four: energy, mobility, water, and sociodemographic data (Rennes Métropole, 2022).

Finally, 3DExperienCity, the third project, consists of the creation of a "metropolitan digital twin." This project aims to predict and anticipate the digital transition in public services, develop new service projects, and promote local economic initiatives. In order to best create new services, such a 3D representation of the city was, finally, enriched with numerous data pertaining to demographics, mobility, urban soil, energy, and vegetation (Rennes Métropole, 2022).

The city of Cardiff in order to best guide its smart city strategies has designed what is known as "Cardiff's Smart City Roadmap." It is a non-mandatory document in which the city desires to become smarter and all the steps to achieve them are included. The strategies in this roadmap are as follows: Capital Ambition, Cardiff Digital Strategy, Cardiff Well-Being Plan, Cardiff Transport Strategy, Cardiff Economic Strategy, and Adopted Local Development Plan (LDP) (Cardiff Council's Smart City Road Map, 2021). To implement the following strategies, the Welsh city will take advantage of the opportunities provided by public services, modernization of public transport, energy infrastructure, and the economic and social benefits caused by digitization and modernization of the urban environment. With regard to public services, Cardiff's goal is to improve their effectiveness and efficiency through the implementation of digital technology solutions. Moreover, the implementation of such solutions will also affect urban mobility. Such implementation will aim to reduce the travel time of trips and, above all, to reduce the environmental impact of public transport (Cardiff Council's Smart City Road Map, 2021). In fact, reducing carbon emissions is one of the key goals of the Welsh city. For this reason, there is an intention to improve the energy infrastructure so that the public sector is energy neutral by 2030. Finally, all these actions aim to cause numerous social and economic effects on the citizens. In fact, digitization and connectivity could help increase citizen productivity and thus lead to economic growth. This would result in more jobs, the creation of social networks, and an acceleration in innovation (Cardiff Council's Smart City Road Map, 2021).

The city of Genève has also embarked on a strategy to increasingly become a smart city. The Genève strategy is based on the concept that innovation depends heavily on citizen involvement: only with the help of the citizenry does it become possible to profoundly transform different territorial specificities. In fact, a city cannot decrease its environmental impact without the help of civil society and various external stakeholders. Therefore, the vision of the Swiss city strategy is to innovate through strong public-private collaborations and partnerships (Smart Geneva, 2022). There are five main areas of focus for the city to do this: mobility, population, government, economy, and environment (Smart Geneva, 2022). Regarding mobility at the center of this focus, the city aims to make the entire transportation network sustainable, collective, collaborative, or alternative. As for population-focused strategies, on the other hand, this area encompasses all projects aimed at improving health services, urban safety, education, cultural development, and local creativity. The government area, as in the case of the other smart cities analyzed so far, aims to put the citizen at the center of decision-making and create a city that is increasingly transparent and accessible to all. Moreover, as anticipated above, the goal of the city of Genève's strategy is to create strong public-private partnerships. This goal is perfectly encapsulated in the "economy" focus area. Indeed, the Swiss city hopes that new jobs can be created through the creation of partnerships. Finally, all the focus areas examined must act on the principle of environmental protection. For this reason, the last area, environment, focuses on creating solutions with low environmental impact and the use of renewable and especially indigenous energy (Smart Geneva, 2022).

The ninth-ranked city, Graz, has undertaken strategies similar to the cities already analyzed. In fact, its "Smart City Graz 2050" strategy is based on several focus areas: society, energy, ecology, mobility, economy, and waste management. Therefore, the goal of the strategy is to create a smart and above all sustainable city in which the well-being of the citizenry is placed at the center. For this reason, the focal areas of the measure are to ensure effective, efficient, and sustainable urban mobility, efficient public services (such as health and education), and above all to reduce its environmental impact by 2050 (Smart City Graz, 2017).

Finally, the tenth-ranked city, Helsinki, has focused its strategies in recent years on accessibility and sustainability. Like many other cities, it has set a goal of energy neutrality by 2035. It is, in fact, one of three flagship cities in the European project "mySMARTlife." This EU-funded project aims to help three cities, Nantes, Hamburg, and Helsinki, be more environmentally friendly through investments and projects focused on reducing CO2 emissions and using renewable energy sources. This project focuses on creating social cohesion, people, and the economy. In fact, the primary goal is not to make these cities technological but rather to ensure that the population has a socially cohesive city that is attractive from a labor and economic perspective but above all, sustainable and environmentally friendly (MySMARTlife, 2022). For this reason, precisely because of the centrality held by the citizenry, all projects and solutions adopted have been co-created together with residents (Helsinki Smart Region, 2022).

At this point, to show the impact of the different dimensions of smart cities individually, we decided to show through Tables 4.5,b,c, and d the results obtained with the corresponding rank. Regarding the first dimension, ICT and transport, the first position is held by Zurich (AR = 1). It is followed in the top 10 by Rotterdam (AR = 2.75), Dresden (AR = 3. 5), Wien (AR = 4), Genève (AR = 4), Amsterdam (AR = 7), Luxembourg (AR = 9), Helsinki (AR = 9), Aalborg (AR = 9.75), and Stockholm (AR = 10.75).

The city of Rotterdam has invested heavily in a project in ICT and transport: the Rotterdam Innovation District. This district is home to numerous companies within it that have made sustainability the focus of their business. In addition, many entrepreneurs in this district can meet and merge their start-ups in such a way as to create more and more innovative solutions. Another ambitious project of the city of Rotterdam is undoubtedly the Rotterdam Smart City Planner, an ad hoc department for urban planning. This body aims to collect a large amount of data aimed at solving different city problems most effectively and efficiently possible (The Global Smart City Knowledge Center, 2020).

The city of Dresden, on the other hand, has decided to pursue the strategy of cooperation to become smarter in ICT and transportation. That is why it has been cooperating with numerous partners, including Volkswagen, in electromobility and digitization projects since 2016. Currently, the VAMOS project is underway, which at its core includes various transport management and control systems for both the city road network and highways. As part of this project, the German Federal Ministry of Transport and Digital Infrastructure (BMVI) is funding research projects on these issues, including trials on driverless cars (TU Dresden, 2021).

Amsterdam, like Dresden, is also pursuing the strategy of collaboration. For this reason, it has fostered partnerships with more than 70 public and private entities. These collaborations have resulted in numerous projects and applications, including Ridesharing, Toogethr Cycles, Toogethr Smart Parking, and Toogethr Rideshare. These projects and applications aim to solve the growing critical transportation issues so that workplaces can be kept accessible and the flow of city traffic can be improved. However, the most ambitious ICT and transportation project in Amsterdam is the Smart Urban Isle. This project aims to increase urban energy savings. A new urban planning concept has been introduced to do this: cities grow and organize themselves based on small integrated areas. For this reason, bioclimatic designs are underway to give each building all sorts of amenities and at the same time energy autonomy at the lowest possible environmental cost (Wilab, 2022).

Finally, the city of Stockholm is investing large sums in the ICT and transportation dimension toward full fiber network coverage. In 2017 it formalized its strategy based on collaboration between civil society, stakeholders, and institutions. This strategy aims to create a communications infrastructure that is completely neutral from competition to meet the diverse needs of citizens and benefit the city's economy (City of Stockholm, 2022).

As done for the ICT and transportation dimensions, we proceed with the analysis of the "People" dimension. The most virtuous cities in this dimension are Groningen (AR = 1.25), Cardiff (AR = 3.25), Aalborg (AR = 3.75), Zürich (AR = 4. 25), Greater Rotterdam (AR = 4.75), Rennes (AR = 6.25), Glasgow City (AR = 7.25), Antwerpen (AR = 9.25), and Oulu (AR = 9.5).

The city of Glasgow has invested heavily in analyzing citizen-generated data in a way that generates value for the city and people first and foremost. In this direction, it has developed the OPEN Data platform through which it has been possible for the city and its organizations to automate the publication of its data, store it easily, and make it widely available (Glasgow City Centre Strategy, 2016).

The city of Antwerpen has also placed the well-being of its citizens at the center of its policies. Particular attention has been paid to mobility. In fact, as city traffic increases, the streets are showing more and more problems in handling a large number of cars. For this reason, the city government has decided to invest in increasingly smart modes of transportation and encourage the citizenry to use railways (The Global Smart City Knowledge Center, 2018).

However, at the heart of Oulu's policies are anthropocentric digital solutions. For this reason, projects for social inclusion and the enhancement of individual soft skills have been implemented. The result of this focus on people's well-being was Oulu's nomination as European Capital of Culture for 2026 (Smart City Oulu, 2020).

In contrast to previous cities, Belfast is investing more on the policy side to improve the privacy and cybersecurity of citizens' personal data. For this reason, it has joined 35 other cities around the world in creating a global policy roadmap for the adoption of new technologies. The ultimate goal is to create close collaboration between business, academia, and the public sector in creating smart city services in key city sectors such as mobility, health, energy, and tourism (Smart Belfast, 2018).

Regarding the size of sustainability below are the smartest cities: Zürich (AR = 1), Rennes (AR = 3.25), Luxembourg (AR = 3.5), Wien (AR = 4. 75), Oulu (AR = 4.75), Bialystok (AR = 5), Malmö (AR = 6), Groningen (AR = 8.5), and München (AR = 11.25).

Bialystok, investing in sustainable policies, has undergone an extremely dynamic transformation to become more and more attractive for business and social capital. It is, in fact, a green city with a strong focus on the physical and mental well-being of its citizens and offers high employment and diverse career opportunities (Intelligent Cities Challenge, 2022).

Malmö, on the other hand, has been awarded numerous times for its brownfield redevelopment projects. Indeed, thanks to this redevelopment work, the city has been able to achieve energy-efficient smart city districts, an effective and efficient sustainable transportation structure, and a seamless implementation of the circular economy. The ultimate goal is to make the entire city grounds energy self-sufficient through renewable energy (Malmö Stad, 2021).

Finally, München has set a 2050 goal of carbon neutrality. Over the years, the city has achieved a great deal of progress. Its infrastructure, in fact, can provide enough electricity from renewable sources to meet the energy needs of the entire city. In addition, München is investing in new renewable energy sources (Landeshauptstadt München, 2021).

Finally, in terms of government size, the most virtuous cities are Zürich (AR = 1), Luxembourg (AR = 2.5), Wien (AR = 3.25), Graz (AR = 3.25), Aalborg (AR = 5. 5), Düsseldorf (AR = 6.75), København (AR = 7.75), Groningen (AR = 8.25), and Rennes (AR = 8.5).

Düsseldorf is investing significantly in e-government. For this reason, it has created a platform where it has grouped all the services of the administration making them free and accessible to all citizens. In addition, the service allows people to download forms, make appointments, and retrieve information quickly and smartly via smartphones as well (Attraktive Verwaltung, 2022).

Finally, as far as København is concerned, it has placed citizenship at the center of policy decisions. It is why the views of civil society are put at the center in the process of developing solutions and why one of the founding values of city policies is accountability (Copenhagen Capacity, 2022).

4.5 Conclusions

This chapter aims to provide different methods to measure smart sustainable city domains. The author chooses four indexes that differ in terms of characteristics and provide construction procedures for each one. They apply both aggregative and non-aggregative methods and compare the obtained results. It emerges that, despite different characteristics and construction procedure are associated with each index, all indexes are highly correlated. It means that the different rankings reflect reciprocally. The application of alternative methods jointly with the results obtained confirms the robustness of the analysis. Finally, this analysis remarks the relevance of adopting an approach that aligns the theory to the methods. The choice of the method should depend on the theoretical assumptions and ensure consistency along the entire research project.

Zurich
Country: Switzerland
Inhabitants: 415,215
Density: 4700/km^2 (12,000/sq. mi)
Total area: 87.88 km^2 (33.93 sq. mi)

The findings in this chapter demonstrate that Zurich, an international banking and financial hub in north Switzerland, is one of the most advanced cities in Europe in terms of governance, sustainability, and ICT. The "Smart City Zurich" concept outlines the growth of the Swiss city and proposes Zurich as a Smart City of global importance by incorporating the initiatives

(continued)

and policies outlined in the prior document, "Strategies Zurich 2035." One of the pillars of this strategy is the digital transformation of municipal services, made possible by collecting data via sensors. This process improves residents' quality of life, protects green spaces in cities, and draws in foreign investment.

The city of Zurich's program is based on four main actions: first, the administration believes that in order to prepare a suitable offer, a thorough ex ante analysis of the population's actual needs, with particular reference to primary services (health, education, and mobility), is necessary.

The collaboration and exchange of ideas among all parties concerned (families, businesses, institutions, etc.), made possible by modern technology and the common use of urban spaces, is another important component of this initiative. In this way, the Smart City Lab has been created, where a selection of solutions are evaluated in collaboration with national and international private stakeholders: data collected by the sensors applied to the city infrastructures are gathered and shared.

The data collection and its free accessibility to the public perfectly encapsulate the third pillar of the Swiss city's strategy. The goal to establish an environment, which is accessible to technological exploration, in order to find the best solutions for the urban fabric of the city, is the fourth and last component of "Smart City Zurich." The local government wants to create cutting-edge, environmentally friendly and shared forms of transportation in the near future: electric busses will take the place of gasoline-powered ones, the current sharing-mobility program will be strengthened, and driverless vehicles' pilots will be started.

Additionally, digitization has an impact on city administration and enables residents to connect online and undertake an increasing number of services; furthermore, innovation is promoted through various instruments, such as loans granted to innovative start-ups to finance their own projects.

A program has also been launched aimed at training administrative staff in ICT matters through collaborations with experts who spend a specific period of time working at the city council, spreading their knowledge to municipal personnel.

In conclusion, the city of Zurich intends to rely mainly on the digital transformation of its services to become a real Smart City of the future.

Appendix

Table 4.5 Index coefficients and ranks for ICT dimension

City	DP2	DP2 ranks	MPI	MPI ranks	SS	SS ranks	AH	AH ranks	AR
Aalborg	16.955	11	107.220	9	6.570	10	68.780	9	10
Ankara	10.593	72	93.696	68	−5.049	69	20.954	70	70
Antalya	9.370	78	90.401	80	−8.275	80	13.253	78	79
Antwerpen	12.712	59	98.107	56	−1.510	56	26.645	63	59
Athina (greater city)	10.453	75	91.641	76	−7.095	76	13.135	80	77
Barcelona	14.785	38	102.193	35	2.146	36	56.535	28	34
Belfast	16.185	21	105.105	19	4.710	20	60.308	19	20
Berlin	16.552	15	103.407	27	3.456	27	56.986	27	24
Bialystok	14.556	42	101.373	40	1.644	41	54.101	33	39
Bologna	12.244	63	96.510	62	−2.501	61	41.264	46	58
Bordeaux	15.350	29	101.870	37	1.803	40	48.207	39	36
Braga	10.839	71	93.385	69	−5.746	73	16.504	76	72
Bratislava	12.925	56	96.833	61	−2.377	60	39.039	49	56
Bruxelles/ Brussels	13.290	51	98.698	54	−0.961	55	33.589	54	54
Bucuresti	10.517	73	92.488	75	−5.946	74	24.798	68	73
Budapest	12.016	65	95.675	65	−3.771	66	20.367	71	67
Burgas	11.030	68	93.023	71	−5.202	70	27.341	61	68
Cardiff	16.176	22	105.048	21	4.656	21	59.910	20	21
Cluj-Napoca	14.457	43	101.151	41	1.922	38	61.854	16	35
Diyarbakir	10.927	70	93.031	70	−5.511	71	30.441	59	68
Dortmund	15.289	31	102.733	32	2.742	33	40.985	47	36
Dresden	18.146	4	109.437	4	8.666	4	79.902	2	4
Dublin (greater city)	13.779	49	100.068	49	0.439	49	44.805	43	48
Düsseldorf	15.208	34	102.938	30	2.928	30	55.574	30	31
Essen	13.944	48	100.708	45	1.054	45	26.746	62	50
Frankfurt am Main	14.768	39	102.482	34	2.761	32	51.537	35	35
Gdansk	14.787	37	100.975	43	1.420	43	46.105	40	41
Genève (greater city)	18.050	5	109.573	3	8.734	3	74.664	5	4
Glasgow City	15.288	32	102.133	36	2.214	35	45.585	41	36
Graz	16.203	20	104.488	24	4.132	24	58.064	25	23
Greater Amsterdam	17.417	9	109.054	6	8.558	5	69.132	8	7

(continued)

Table 4.5 (continued)

City	DP2	DP2 ranks	MPI	MPI ranks	SS	SS ranks	AH	AH ranks	AR
Greater Manchester	16.157	23	104.695	22	4.393	22	48.637	38	26
Greater Rotterdam	18.432	3	110.251	2	9.465	2	75.071	4	3
Groningen	16.493	16	105.806	14	5.713	14	49.284	37	20
Hamburg	16.231	19	105.161	17	4.954	16	66.651	10	16
Helsinki/ Helsingfors (greater city)	17.599	8	108.653	8	8.012	8	65.114	12	9
Irakleio	9.123	79	90.644	79	−7.917	79	11.612	81	80
Istanbul	9.098	80	90.735	78	−7.876	78	13.140	79	79
København (greater city)	16.271	18	106.851	11	6.400	11	61.445	18	15
Köln	12.756	58	98.229	55	−0.914	54	19.496	74	60
Kosice	13.273	52	98.710	53	−0.774	53	39.039	49	52
Kraków	12.213	64	97.273	59	−1.970	59	25.984	66	62
Lefkosia	7.992	81	86.111	82	−11.537	81	15.297	77	80
Leipzig	14.170	45	101.442	39	1.860	39	37.329	51	44
Liège	12.399	61	96.952	60	−2.575	62	26.168	65	62
Lille	14.101	46	99.195	51	−0.473	51	32.718	56	51
Lisboa (greater city)	10.494	74	90.801	77	−7.816	77	19.712	72	75
Ljubljana	14.701	40	101.041	42	1.148	44	55.409	31	39
London (greater city)	16.313	17	106.030	13	5.776	13	59.514	22	16
Luxembourg	17.676	6	107.070	10	6.739	9	65.843	11	9
Madrid	15.010	35	102.759	31	2.601	34	59.640	21	30
Málaga	15.309	30	101.831	38	1.950	37	54.287	32	34
Malmö	16.081	24	105.154	18	4.754	19	64.779	13	19
Marseille	13.074	55	97.649	58	−1.953	58	22.947	69	60
Miskolc	12.868	57	99.009	52	−0.669	52	42.758	44	51
München	15.955	26	104.497	23	4.338	23	56.196	29	25
Napoli (greater city)	3.354	84	72.798	85	−22.710	85	4.058	85	85
Narva	12.535	60	96.256	63	−2.862	63	30.966	57	61
Oslo	17.641	7	108.800	7	8.441	7	59.097	23	11
Ostrava	16.007	25	103.148	29	3.237	28	49.300	36	30
Oulu	13.536	50	100.727	44	1.521	42	32.941	55	48
Oviedo	15.261	33	103.251	28	3.038	29	57.002	26	29
Palermo	2.907	85	73.801	83	−22.230	84	4.384	84	84
Paris (greater city)	14.251	44	100.482	47	0.539	48	37.196	52	48
Piatra Neamt	11.542	67	95.298	66	−3.620	65	30.554	58	64

(continued)

Table 4.5 (continued)

City	DP2	DP2 ranks	MPI	MPI ranks	SS	SS ranks	AH	AH ranks	AR
Praha	17.214	10	105.522	15	5.462	15	64.227	14	14
Rennes	15.768	27	103.719	25	3.458	26	61.619	17	24
Reykjavík	11.696	66	93.005	72	−3.911	67	45.120	42	62
Riga	10.105	77	93.730	67	−4.736	68	19.603	73	71
Roma	4.123	83	73.304	84	−21.654	83	6.281	82	83
Sofia	10.934	69	92.586	74	−5.730	72	24.887	67	71
Stockholm (greater city)	16.725	12	106.367	12	5.845	12	71.314	7	11
Strasbourg	15.617	28	103.704	26	3.497	25	61.946	15	24
Stuttgart	14.627	41	102.519	33	2.765	31	41.807	45	38
Tallinn	16.681	13	105.095	20	4.832	18	72.005	6	14
Tartu	13.215	53	98.001	57	−1.661	57	26.364	64	58
Torino	7.956	82	86.870	81	−11.606	82	4.779	83	82
Tyneside conurbation	16.676	14	105.250	16	4.859	17	58.434	24	18
Valletta	12.263	62	95.980	64	−2.968	64	34.805	53	61
Verona	10.378	76	92.920	73	−6.117	75	18.335	75	75
Vilnius	13.088	54	99.297	50	−0.268	50	29.896	60	54
Warszawa	14.095	47	100.699	46	0.959	46	53.173	34	43
Wien	18.795	2	109.340	5	8.527	6	77.417	3	4
Zagreb	14.885	36	100.363	48	0.851	47	38.733	50	45
Zürich (greater city)	20.432	1	113.229	1	12.002	1	83.577	1	1

DP2 P2 distance; *MPI* Mazziotta-Pareto index; *SS* sum of standardized indicators; *AH* average height; *AR* average rank between DP2, MPI, SS, AH ranks

Table 4.6 Index coefficients and ranks for people dimension

City	DP2	DP2 ranks	MPI	MPI ranks	SS	SS ranks	AH	AH ranks	AR
Aalborg	25.726	5	108.724	3	12.492	3	68.730	4	4
Ankara	16.757	70	94.092	69	−7.563	70	42.998	46	64
Antalya	18.045	66	95.610	65	−5.019	65	42.998	46	61
Antwerpen	24.476	10	105.807	11	8.629	10	68.081	6	9
Athina (greater city)	6.902	85	78.929	85	−26.608	85	26.520	75	83
Barcelona	18.364	63	96.910	63	−3.696	63	30.260	70	65
Belfast	24.209	12	105.890	10	8.484	11	67.492	7	10
Berlin	18.511	62	97.148	61	−3.408	61	18.548	81	66
Bialystok	23.607	17	103.341	26	5.828	27	42.998	46	29
Bologna	19.089	58	97.588	57	−2.995	60	27.053	73	62
Bordeaux	23.478	18	104.313	19	6.373	21	50.301	24	21
Braga	23.336	21	103.276	27	5.621	28	54.186	15	23
Bratislava	16.089	71	93.107	72	−8.720	72	42.998	46	65
Bruxelles/Brussels	18.916	60	97.143	62	−3.583	62	30.821	69	63
Bucuresti	14.291	77	90.017	77	−12.832	77	34.649	64	74
Budapest	17.681	67	94.109	68	−7.301	68	34.649	64	67
Burgas	21.171	46	99.463	49	0.009	49	35.249	62	52
Cardiff	25.861	3	107.725	4	10.941	4	70.336	2	3
Cluj-Napoca	23.170	24	103.912	24	6.075	25	46.958	28	25
Diyarbakir	18.109	64	94.894	67	−5.723	66	42.998	46	61
Dortmund	19.869	53	99.098	51	−0.624	51	46.958	28	46
Dresden	23.074	26	104.034	23	6.321	22	52.271	21	23
Dublin (greater city)	22.265	36	102.828	33	4.507	34	42.998	46	37
Düsseldorf	23.099	25	104.303	20	6.866	20	42.998	46	28
Essen	19.231	56	98.373	53	−1.617	53	42.998	46	52

(continued)

Table 4.6 (continued)

City	DP2	DP2 ranks	MPI	MPI ranks	SS	SS ranks	AH	AH ranks	AR
Frankfurt am Main	21.201	45	101.072	43	2.729	41	42.998	46	44
Gdansk	22.938	27	102.946	31	4.703	33	53.593	16	27
Genève (greater city)	24.397	11	104.932	16	7.676	16	42.998	46	22
Glasgow City	24.993	8	106.975	6	9.907	6	62.117	9	7
Graz	23.767	14	105.716	12	8.474	12	56.984	10	12
Greater Amsterdam	25.143	6	106.577	8	9.887	7	42.998	46	17
Greater Manchester	22.329	35	102.977	30	4.263	36	53.069	18	30
Greater Rotterdam	25.858	4	107.344	5	10.719	5	68.710	5	5
Groningen	26.746	1	109.208	1	13.266	2	70.600	1	1
Hamburg	21.284	44	101.653	38	3.093	38	42.998	46	42
Helsinki/Helsingfors (greater city)	23.610	16	104.995	15	7.715	15	42.998	46	23
Irakleio	13.172	79	88.561	79	−14.168	79	30.976	68	76
Istanbul	10.594	83	83.999	83	−20.518	83	20.417	80	82
København (greater city)	22.701	30	103.747	25	5.955	26	42.998	46	32
Köln	20.150	49	100.589	45	1.650	45	42.998	46	46
Kosice	20.067	51	98.146	55	−2.168	56	30.151	71	58
Kraków	22.766	29	102.430	36	3.916	37	54.393	14	29
Lefkosia	15.889	72	93.390	71	−8.408	71	29.488	72	72
Leipzig	21.467	41	101.380	40	2.692	42	46.958	28	38
Liège	18.079	65	95.827	64	−4.881	64	33.964	65	65
Lille	19.433	55	98.187	54	−1.628	54	42.998	46	52
Lisboa (greater city)	15.526	73	92.491	73	−9.659	73	14.787	84	76
Ljubljana	21.464	42	101.470	39	2.783	40	39.213	61	46
London (greater city)	21.538	40	101.312	41	2.167	43	42.998	46	43
Luxembourg	24.019	13	105.475	13	8.339	13	42.998	46	21

Madrid	13.331	78	89.455	78	−13.187	78	18.274	82	79
Málaga	20.104	50	99.162	50	−0.032	50	52.272	19	42
Malmö	22.863	28	102.579	34	4.811	30	46.958	28	30
Marseille	15.185	75	92.139	74	−10.269	74	23.056	78	75
Miskolc	15.405	74	90.662	76	−12.045	76	25.846	77	76
München	23.421	20	104.867	17	7.611	17	42.998	46	25
Napoli (greater city)	11.944	81	85.994	81	−17.622	81	30.976	68	78
Narva	19.223	57	97.167	60	−2.737	58	42.998	46	55
Oslo	23.240	22	105.223	14	8.034	14	52.271	21	18
Ostrava	21.297	43	100.117	47	1.429	47	46.958	28	41
Oulu	24.762	9	106.240	9	9.648	9	56.831	11	10
Oviedo	22.642	32	102.579	35	4.803	31	51.442	23	30
Palermo	8.930	84	81.458	84	−23.583	84	16.788	83	84
Paris (greater city)	18.958	59	97.533	58	−2.798	59	25.908	76	63
Piatra Neamt	18.904	61	97.345	59	−2.618	57	51.442	23	50
Praha	22.662	31	102.326	37	4.448	35	53.373	17	30
Rennes	25.099	7	106.758	7	9.849	8	69.343	3	6
Reykjavík	20.587	47	101.203	42	2.906	39	42.998	46	44
Riga	17.614	68	94.944	66	−6.401	67	21.127	79	70
Roma	10.973	82	85.841	82	−19.105	82	8.000	85	83
Sofia	12.569	80	86.953	80	−16.484	80	42.998	46	72
Stockholm (greater city)	23.439	19	104.283	21	7.170	18	42.998	46	26
Strasbourg	23.170	23	104.141	22	6.118	23	55.056	13	20
Stuttgart	22.448	34	103.252	28	6.118	24	42.998	46	33
Tallinn	19.873	52	98.723	52	−1.104	52	42.998	46	51
Tartu	22.615	33	102.846	32	4.709	32	46.958	28	31
Torino	17.504	69	94.075	70	−7.550	69	32.322	66	69

(continued)

Table 4.6 (continued)

City	DP2	DP2 ranks	MPI	MPI ranks	SS	SS ranks	AH	AH ranks	AR
Tyneside conurbation	23.620	15	104.839	18	7.093	19	64.756	8	15
Valletta	14.456	76	90.802	75	−10.426	75	26.636	74	75
Verona	20.429	48	100.073	48	0.378	48	46.958	28	43
Vilnius	21.776	38	100.578	46	1.568	46	42.998	46	44
Warszawa	19.494	54	97.938	56	−1.920	55	42.998	46	53
Wien	22.116	37	103.042	29	5.115	29	42.998	46	35
Zagreb	21.647	39	100.675	44	1.678	44	42.998	46	43
Zürich (greater city)	26.275	2	109.018	2	13.435	1	56.009	12	4

DP2 P2 distance; *MPI* Mazziotta-Pareto index; *SS* sum of standardized indicators; *AH* average height; *AR* average rank between DP2, MPI, SS, AH ranks

Table 4.7 Index coefficients and ranks for sustainability dimension

City	DP2	DP2 ranks	MPI	MPI ranks	SS	SS ranks	AH	AH ranks	AR
Aalborg	9.399	16	107.024	19	3.757	15	66.913	16	17
Ankara	6.992	50	100.348	47	0.264	47	33.223	51	49
Antalya	6.241	57	97.396	56	−1.235	56	30.241	57	57
Antwerpen	8.269	36	103.476	38	1.770	40	49.732	40	39
Athina (greater city)	1.360	84	78.187	84	−10.646	84	3.921	85	84
Barcelona	4.813	73	91.996	71	−3.820	72	18.064	71	72
Belfast	9.546	13	107.711	13	3.905	13	67.859	15	14
Berlin	6.572	54	98.625	50	−0.545	51	35.402	49	51
Bialystok	10.019	7	110.452	5	5.277	5	78.569	3	5
Bologna	6.202	59	96.283	59	−1.735	60	31.110	56	59
Bordeaux	9.590	11	108.050	12	4.289	11	65.138	19	13
Braga	8.711	30	104.924	31	2.603	31	55.198	32	31
Bratislava	3.959	75	86.529	76	−6.436	76	9.519	78	76
Bruxelles/Brussels	5.431	67	94.025	66	−2.836	66	21.661	67	67
Bucuresti	2.276	82	82.854	81	−8.385	81	5.031	81	81
Budapest	5.026	70	92.616	69	−3.615	70	17.294	72	70
Burgas	6.775	52	97.492	55	−0.767	53	38.089	45	51
Cardiff	9.521	14	108.482	11	4.259	12	74.282	11	12
Cluj-Napoca	5.750	63	95.436	61	−2.117	61	27.504	59	61
Diyarbakir	6.496	56	98.315	52	−0.667	52	37.561	46	52
Dortmund	8.108	39	103.522	36	1.869	37	55.850	31	36
Dresden	8.751	28	105.981	24	3.019	25	58.872	28	26
Dublin (greater city)	8.831	26	104.240	33	2.423	33	52.526	37	32
Düsseldorf	8.529	35	105.214	28	2.620	30	56.003	30	31
Essen	7.484	44	101.059	44	0.656	44	38.131	44	44
Frankfurt am Main	6.837	51	99.429	49	−0.259	49	35.613	48	49
Gdansk	7.547	43	101.993	43	1.105	43	41.379	43	43

(continued)

Table 4.7 (continued)

City	DP2	DP2 ranks	MPI	MPI ranks	SS	SS ranks	AH	AH ranks	AR
Genève (greater city)	8.674	32	106.079	23	3.095	23	65.152	18	24
Glasgow City	9.590	10	107.191	17	3.691	18	63.835	22	17
Graz	8.698	31	103.406	40	1.923	36	53.538	35	36
Greater Amsterdam	8.053	41	103.445	39	1.771	39	48.430	41	40
Greater Manchester	9.062	20	105.138	30	2.723	29	59.531	25	26
Greater Rotterdam	8.571	33	103.998	35	2.047	35	52.421	39	36
Groningen	9.851	8	109.442	9	4.729	9	76.228	8	9
Hamburg	8.539	34	105.687	27	2.860	28	56.484	29	30
Helsinki/Helsingfors (greater city)	9.823	9	109.592	8	4.830	8	76.038	9	9
Irakleio	3.463	78	85.208	78	−6.588	77	19.284	69	76
Istanbul	3.486	77	88.206	75	−5.591	75	14.135	76	76
København (greater city)	8.079	40	103.493	37	1.777	38	52.965	36	38
Köln	6.739	53	98.005	54	−0.934	55	31.156	55	54
Kosice	6.138	61	96.550	57	−1.707	57	26.075	61	59
Kraków	5.389	68	92.150	70	−3.387	69	16.715	73	70
Lefkosia	5.981	62	94.209	65	−2.666	65	21.927	64	64
Leipzig	8.734	29	106.634	22	3.357	22	64.897	21	24
Liège	6.193	60	95.184	62	−2.311	62	27.112	60	61
Lille	7.474	45	100.538	46	0.342	46	35.260	50	47
Lisboa (greater city)	4.851	72	91.873	73	−4.036	73	14.731	74	73
Ljubljana	9.019	22	107.231	16	3.654	19	60.144	24	20
London (greater city)	8.123	38	102.223	42	1.245	41	52.460	38	40
Luxembourg	10.224	4	111.161	2	5.643	2	77.649	6	4
Madrid	3.152	79	85.076	79	−7.302	79	5.762	79	79
Málaga	5.480	66	94.256	64	−2.633	64	21.729	65	65

Malmö	10.204	5	110.103	7	5.130	7	77.844	5	6
Marseille	4.478	74	89.096	74	−5.304	74	14.221	75	74
Miskolc	7.000	49	98.008	53	−0.877	54	33.217	52	52
München	9.464	15	109.043	10	4.550	10	75.150	10	11
Napoli (greater city)	2.067	83	79.827	83	−9.804	83	4.193	84	83
Narva	8.852	24	104.853	32	2.550	32	54.847	33	30
Oslo	9.031	21	106.706	21	3.376	21	69.971	12	19
Ostrava	6.522	55	95.760	60	−1.713	59	24.930	62	59
Oulu	10.236	3	110.347	6	5.241	6	77.917	4	5
Oviedo	8.833	25	105.847	25	3.061	24	59.166	26	25
Palermo	0.604	85	75.805	85	−11.885	85	4.369	83	85
Paris (greater city)	5.099	69	91.982	72	−3.704	71	19.173	70	71
Piatra Neamt	9.132	19	106.951	20	3.533	20	58.904	27	22
Praha	5.578	65	94.757	63	−2.482	63	19.301	68	65
Rennes	10.330	2	110.920	4	5.518	4	79.164	2	3
Reykjavík	7.384	46	99.742	48	0.235	48	32.431	53	49
Riga	7.065	48	98.335	51	−0.282	50	32.331	54	51
Roma	2.281	81	81.004	82	−9.255	82	4.858	82	82
Sofia	2.977	80	85.057	80	−7.345	80	5.306	80	80
Stockholm (greater city)	9.252	17	107.391	15	3.710	16	68.784	14	16
Strasbourg	8.915	23	105.180	29	3.012	26	65.092	20	25
Stuttgart	7.969	42	102.398	41	1.221	42	46.535	42	42
Tallinn	8.190	37	104.130	34	2.138	34	54.201	34	35
Tartu	9.146	18	107.138	18	3.707	17	66.000	17	18
Torino	5.689	64	93.719	67	−2.902	67	21.721	66	66
Tyneside conurbation	9.570	12	107.551	14	3.857	14	69.767	13	13
Valletta	3.496	76	85.768	77	−6.915	78	11.367	77	77

(continued)

Table 4.7 (continued)

City	DP2	DP2 ranks	MPI	MPI ranks	SS	SS ranks	AH	AH ranks	AR
Verona	6.218	58	96.421	58	−1.711	58	29.499	58	58
Vilnius	8.827	27	105.791	26	2.903	27	60.973	23	26
Warszawa	5.006	71	93.206	68	−3.094	68	23.193	63	68
Wien	10.086	6	111.062	3	5.588	3	76.680	7	5
Zagreb	7.346	47	100.851	45	0.464	45	35.884	47	46
Zürich (greater city)	10.410	1	112.310	1	6.196	1	79.662	1	1

DP2 P2 distance; *MPI* Mazziotta-Pareto index; *SS* sum of standardized indicators; *AH* average height; *AR* average rank between DP2, MPI, SS, AH ranks

Table 4.8 Index coefficients and ranks for government dimension

City	DP2	DP2 ranks	MPI	MPI ranks	SS	SS ranks	AH	AH ranks	AR
Aalborg	17.633	5	110.290	5	8.709	5	68.328	7	6
Ankara	15.863	11	107.140	11	6.048	11	67.403	9	11
Antalya	13.287	36	102.381	34	2.279	34	32.429	59	41
Antwerpen	13.665	32	102.685	32	2.880	27	46.533	41	33
Athina (greater city)	4.809	82	83.320	82	−12.562	81	4.658	85	83
Barcelona	11.549	56	98.227	53	−1.160	54	28.171	66	57
Belfast	13.204	37	103.121	28	2.708	29	53.055	28	31
Berlin	9.440	72	92.021	70	−5.685	72	27.302	67	70
Bialystok	15.510	13	105.491	16	4.557	16	62.771	12	14
Bologna	13.498	33	101.995	37	1.722	37	46.648	40	37
Bordeaux	14.569	22	104.765	19	3.863	22	52.489	29	23
Braga	13.189	38	102.500	33	2.269	35	52.347	30	34
Bratislava	9.298	73	92.119	69	−5.702	73	23.282	71	72
Bruxelles/Brussels	14.933	18	105.557	15	4.871	15	56.792	21	17
Bucuresti	8.289	75	91.265	73	−6.400	74	16.920	75	74
Budapest	11.503	58	99.097	49	−0.596	51	24.300	69	57
Burgas	11.332	60	98.536	51	−0.412	49	37.208	56	54
Cardiff	15.706	12	107.647	10	6.178	10	63.836	11	11
Cluj-Napoca	12.422	47	100.295	43	1.021	40	48.213	36	42
Diyarbakir	11.102	63	97.987	56	−1.237	58	33.044	58	59
Dortmund	12.457	46	98.943	50	−0.547	50	47.072	39	46
Dresden	12.258	51	97.358	61	−1.179	56	39.350	53	55
Dublin (greater city)	14.520	25	104.204	23	3.515	23	53.423	27	25
Düsseldorf	16.468	7	108.374	7	6.814	7	74.956	6	7
Essen	13.151	39	99.327	48	−0.098	48	51.153	33	42

(continued)

Table 4.8 (continued)

City	DP2	DP2 ranks	MPI	MPI ranks	SS	SS ranks	AH	AH ranks	AR
Frankfurt am Main	15.261	17	104.747	20	3.999	19	57.444	20	19
Gdansk	14.541	24	103.695	25	3.191	25	60.754	15	22
Genève (greater city)	15.994	9	106.795	12	5.755	12	68.213	8	10
Glasgow City	13.467	34	102.779	30	2.403	33	47.947	37	34
Graz	18.037	4	112.398	3	9.938	3	81.467	3	3
Greater Amsterdam	10.939	65	91.894	71	−4.623	69	31.008	61	67
Greater Manchester	15.342	15	106.748	13	5.608	13	60.620	17	15
Greater Rotterdam	10.758	67	91.777	72	−4.810	70	22.671	73	71
Groningen	15.902	10	107.748	9	6.399	9	75.019	5	8
Hamburg	14.665	20	103.131	27	2.697	30	51.289	32	27
Helsinki/Helsingfors (greater city)	13.695	30	102.355	35	2.427	32	56.523	22	30
Irakleio	7.990	77	90.939	74	−6.691	75	13.449	77	76
Istanbul	11.088	64	97.655	60	−1.642	60	24.089	70	64
København (greater city)	16.932	6	108.744	6	7.400	6	62.363	13	8
Köln	10.910	66	90.774	75	−5.296	71	54.585	24	59
Kosice	11.730	54	98.056	55	−1.100	53	29.863	63	56
Kraków	12.877	42	99.866	47	0.033	47	28.432	65	50
Lefkosia	12.336	49	100.069	46	0.634	44	39.298	54	48
Leipzig	13.672	31	101.601	38	1.436	38	42.078	49	39
Liège	14.379	26	104.420	22	3.923	21	51.029	34	26
Lille	12.585	45	100.610	41	0.734	42	29.889	62	48
Lisboa (greater city)	10.045	70	95.506	65	−3.303	65	24.558	68	67
Ljubljana	13.297	35	102.038	36	1.893	36	54.075	26	33
London (greater city)	13.707	29	103.644	26	3.059	26	52.078	31	28
Luxembourg	18.641	2	114.234	2	11.506	2	81.234	4	3

Madrid	12.287	50	98.255	52	−1.163	55	39.992	52	52
Málaga	12.648	44	100.149	45	0.195	46	42.937	46	45
Malmö	14.648	21	103.068	29	2.869	28	47.689	38	29
Marseille	9.866	71	94.116	67	−4.469	68	14.904	76	71
Miskolc	12.714	43	101.092	40	1.009	41	48.227	35	40
München	14.932	19	104.725	21	3.976	20	61.839	14	19
Napoli (greater city)	5.302	81	83.596	81	−12.822	82	5.758	82	82
Narva	4.071	84	79.421	85	−15.716	85	5.337	84	85
Oslo	11.379	59	97.690	59	−0.832	52	44.196	44	54
Ostrava	12.362	48	100.169	44	0.546	45	45.298	42	45
Oulu	11.969	52	92.353	68	−4.086	67	42.878	47	59
Oviedo	11.740	53	97.861	57	−1.388	59	44.654	43	53
Palermo	4.025	85	80.663	84	−14.752	84	5.921	81	84
Paris (greater city)	13.079	40	101.480	39	1.254	39	37.701	55	43
Piatra Neamt	11.188	62	96.860	63	−1.739	62	41.738	50	59
Praha	10.625	68	96.501	64	−2.355	64	43.772	45	60
Rennes	16.144	8	108.039	8	6.571	8	65.657	10	9
Reykjavík	10.106	69	94.780	66	−3.616	66	31.796	60	65
Riga	7.227	78	88.746	77	−8.402	77	8.070	78	78
Roma	4.404	83	82.039	83	−13.724	83	6.020	80	82
Sofia	5.817	80	86.607	79	−9.865	79	7.874	79	79
Stockholm (greater city)	14.024	27	102.743	31	2.623	31	60.692	16	26
Strasbourg	15.267	16	106.060	14	5.035	14	58.203	19	16
Stuttgart	15.499	14	105.050	18	4.255	18	55.755	23	18
Tallinn	12.978	41	100.442	42	0.679	43	41.546	51	44
Tartu	8.078	76	85.769	80	−9.646	78	29.130	64	75
Torino	8.571	74	90.587	76	−6.923	76	18.806	74	75

(continued)

Table 4.8 (continued)

City	DP2	DP2 ranks	MPI	MPI ranks	SS	SS ranks	AH	AH ranks	AR
Tyneside conurbation	13.968	28	103.765	24	3.264	24	54.308	25	25
Valletta	14.563	23	105.305	17	4.292	17	60.057	18	19
Verona	11.257	61	98.087	54	−1.195	57	42.838	48	55
Vilnius	11.659	55	97.743	58	−1.699	61	36.604	57	58
Warszawa	11.505	57	96.869	62	−2.205	63	22.671	72	64
Wien	18.057	3	111.985	4	9.616	4	81.475	2	3
Zagreb	6.114	79	87.051	78	−10.097	80	5.732	83	80
Zürich (greater city)	19.980	1	116.304	1	13.081	1	83.263	1	1

DP2 P2 distance; *MPI* Mazziotta-Pareto index; *SS* sum of standardized indicators; *AH* average height; *AR* average rank between DP2, MPI, SS, AH ranks

References

Alaimo, L. S., Arcagni, A., Fattore, M., & Maggino, F. (2021a). Synthesis of multi-indicator system over time: A poset-based approach. *Social Indicators Research, 157*(1), 77–99.

Alaimo, L. S., Ciacci, A., & Ivaldi, E. (2021b). Measuring sustainable development by non-aggregative approach. *Social Indicators Research, 157*(1), 101–122.

Alaimo, L. S., Ivaldi, E., Landi, S., & Maggino, F. (2022a). Measuring and evaluating socio-economic inequality in small areas: An application to the urban units of the Municipality of Genoa. *Socio-Economic Planning Sciences, 83*, 101170. https://doi.org/10.1016/j.seps.2021.101170

Alaimo, L. S., Arcagni, A., Fattore, M., Maggino, F., & Quondamstefano, V. (2022b). Measuring equitable and sustainable well-being in Italian Regions: The Non-aggregative approach. *Social Indicators Research, 161*(2–3), 711–733. https://doi.org/10.1007/s11205-020-02388-7

Arcagni, A., Barbiano di Belgiojoso, E., Fattore, M., & Rimoldi, S. (2018). Multidimensional analysis of deprivation and fragility patterns of migrants in Lombardy, using partially ordered sets and self-organizing Mapsin. *Social Indicators Research, 141*(2), 551–579.

Attraktive Verwaltung. (2022). *E-Government in Düsseldorf: Serviceportal für Online-Behördengänge › Attraktive Verwaltung*. Accessed Jan 23, 2023, from https://www.attraktive-verwaltung.de/e-government-in-duesseldorf-serviceportal-fuer-online-behoerdengaenge

Bonatti, G., Ciacci, A., & Ivaldi, E. (2021). Different measures of country risk: An application to European countries. *Journal of Risk and Financial Management, 14*(1), 19.

Bruzzi, C., Ivaldi, E., & Landi, S. (2020). Non-compensatory aggregation method to measure social and material deprivation in an urban area: Relationship with premature mortality. *The European Journal of Health Economics, 21*(3), 381–396.

Cardiff Council's Smart City Road Map. (2021). *Smart Cardiff*. Available on: 0000765 RC Smart Cities 2019 FIN craig.indd (smartcardiff.co.uk).

Carlsen, L., & Bruggemann, R. (2017). Partial ordering and metrology analyzing analytical performance. In M. Fattore & R. Bruggemann (Eds.), *Partial order concepts in applied sciences* (pp. 49–70). Springer.

Carstairs, V., & Morris, R. (1991). *Deprivation and Health in Scotland*. Aberdeen University Press.

Ciacci, A., Ivaldi, E., & González-Relaño, R. (2021a). A partially non-compensatory method to measure the smart and sustainable level of Italian municipalities. *Sustainability, 13*(1), 435.

Ciacci, A., Ivaldi, E., Mangano, S., & Ugolini, G. M. (2021b). Environment, logistics and infrastructure: the three dimensions of influence of Italian coastal tourism. *Journal of Sustainable Tourism*, 1–21.

Ciacci, A., Ivaldi, E., & Soliani, R. (2021c). A Potential Business Environment of Smart Cities: A Subjective Approach. In *Strategic Outlook in Business and Finance Innovation: Multidimensional Policies for Emerging Economies*. Emerald Publishing Limited.

City of Stockholm. (2022). *The smart city–city of Stockholm*. Start–City of Stockholm. Accessed Jan 23, 2023, from https://international.stockholm.se/city-development/the-smart-city/

Copenhagen Capacity. (2022). *Smart city in Greater Copenhagen. Invest in Greater Copenhagen*. Accessed Jan 23, 2023, from https://www.copcap.com/set-up-a-business/key-sectors/smart-city

Davey, B. A., & Priestley, B. H. (2002). *Introduction to lattices and order*. Cambridge University Press.

Etienne, S. (2019, September 27). *Luxembourg-Ville, L'exemple Même De La Smart City*. smartcitiesmag.lu. Accessed Jan 23, 2023, from https://smartcitiesmag.lu/web/luxembourg-ville-lexemple-meme-de-la-smart-city/

European Commission. (2022). *Smart cities*. European Commission–European Commission. Accessed Jan 23, 2023, from https://ec.europa.eu/info/eu-regional-and-urban-development/topics/cities-and-urban-development/city-initiatives/smart-cities_en

Fattore, M. (2017a). Synthesis of indicators: The non-aggregative approach. In *Complexity in society: From indicators construction to their synthesis* (pp. 193–212). Springer.

Fattore, M. (2017b). Functionals and synthetic indicators over finite posets. In M. Fattore & R. Bruggemann (Eds.), *Partial order concepts in applied sciences* (pp. 71–86). Cham.

Fattore, M. (2018). Non-aggregated indicators of environmental sustainability. *Silesian Statistical Review/Slaski Przeglad Statystyczny, 16*(22), 7–22.

Fattore, M., & Arcagni, A. (2018). F-FOD: Fuzzy first order dominance analysis and populations ranking over ordinal multi-indicator systems. *Social Indicators Research, 144*(1), 1–29.

Fattore, M., Arcagni, A., & Maggino, F. (2019). Optimal scoring of partially ordered data, with an application to the ranking of smart cities. In G. Arbia, S. Peluso, A. Pini, & G. Rivellini (Eds.), *SIS 2019–smart statistics for smart applications, Pearson, Milano* (pp. 855–860). Società Italiana di Statistica.

Forrest, R., & Gordon, D. (1993). *People and places: A 1991 Census Atlas of England*. University of Bristol, SAUS.

Freudenberg, M. (2003). *Composite indicators of country performance: A critical assessment*. STI Working Paper, 2003/16, Industry Issues. OECD.

Glasgow City Centre Strategy. (2016, September 30). *What are smart cities? | Glasgow City Centre Strategy*. Glasgow City Centre Strategy | creating a green, liveable city centre that fosters creativity and opportunity. Accessed Jan 23, 2023, from https://www.glasgowcitycentrestrategy.com/what-are-smart-cities.htm

Helsinki Smart Region (2022). *Helsinkismart*. Helsinkismart. Accessed Jan 23, 2023, from https://helsinkismart.fi/

Hilckmann, A., Bach, V., Bruggemann, R., Ackermann, R., & Finkbeiner, M. (2017). Partial order analysis of the government dependence of the sustainable development performance in Germany's federal states. In *Partial Order Concepts in Applied Sciences* (pp. 219–228). Springer.

Intelligent Cities Challenge (2022). *Białystok | Intelligent cities challenge*. Home | Intelligent Cities Challenge. Accessed Jan 23, 2023, from https://www.intelligentcitieschallenge.eu/cities/bialystok

Ivaldi, E., Ciacci, A., & Soliani, R. (2020a). Urban deprivation in Argentina: A POSET analysis. *Papers in Regional Science, 99*(6), 1723–1747.

Ivaldi, E., Parra Saiani, P., Primosich, J. J., & Bruzzi, C. (2020b). Health and deprivation: A new approach applied to 32 Argentinian urban areas. *Social Indicators Research, 151*(1), 155–179.

Ivaldi, E., Penco, L., Isola, G., & Musso, E. (2020c). Smart sustainable cities and the urban knowledge-based economy: A NUTS3 level analysis. *Social Indicators Research, 150*(1), 45–72.

Ivaldi, E., & Testi, A. (2011). In C. M. Baird (Ed.), *Genoa Index of Deprivation (GDI): An index of material deprivation for geographical areas in social indicators: Statistics, trends and policy development* (pp. 75–98). Nova Publisher.

Ivanović, B. (1974). A method of establishing a list of development indicators. *Economic Analysis, 8*(1–2), 52–64.

Landeshauptstadt München. (2021). *München als smart city*. Rathaus–Landeshauptstadt München. Accessed Jan 23, 2023, from https://stadt.muenchen.de/infos/muenchen-smart-city.html

Landi, S., Ivaldi, E., & Testi, A. (2018). Measuring change over time in socio-economic deprivation and health in an urban context: The case study of Genoa. *Social Indicators Research, 139*, 745–785.

Maggino, F. (2017a). Developing indicators and managing the complexity. In *Complexity in society: From indicators construction to their synthesis* (pp. 87–114). Springer.

Maggino, F. (2017b). Dealing with syntheses in a system of indicators. In *Complexity in society: From indicators construction to their synthesis* (pp. 115–137). Springer.

Making City. (2019). *GRONINGEN–making city*. Accessed Jan 23, 2023, from https://Makingcity.Eu/Groningen/. https://makingcity.eu/groningen/

Malmö stad. (2021, March 31). *Theme sustainable city*. Startsidan–Malmö stad. Accessed Jan 23, 2023, from https://malmo.se/Welcome-to-Malmo/Technical-visits/Theme-Sustainable-City.html

Marsal-Llacuna, M.-L., Colomer-Llinàs, J., & Meléndez-Frigola, J. (2015). Lessons in urban monitoring taken from sustainable and livable cities to better address the smart cities initiative. *Technological Forecasting and Social Change, 90*, 611–622.

Mazziotta, M., & Pareto, A. (2017). Synthesis of indicators: The composite indicators approach. In *Complexity in society: From indicators construction to their synthesis* (pp. 159–191). Springer.

Mazziotta, M., & Pareto, A. (2020). Composite indices construction: The performance interval approach. *Social Indicators Research*, 1–11.

MySMARTLife. (2022). *MySMARTLife*. Accessed Jan 23, 2023, from https://www.mysmartlife.eu/mysmartlife/

Montero, J.-M., Coro, C., & Beatriz, L. (2010). Building an environmental quality index for a big city: A spatial interpolation approach combined with a distance indicator. *Journal of Geographical Systems, 12*, 435–459.

Nayak, P., & Mishra, S. K. (2012). *Efficiency of Pena's P2 distance in construction of human evelopment indices*. MPRA—Munich Personal RePEc Archive, paper 39022.3

Neggers, J., & Kim, H. S. (1998). *Basic posets*. World Scientific.

Norman, P. (2010). Identifying change over time in small area socio-economic deprivation. *Applied Spatial Analysis and Policy, 3*(2), 107–138.

Pena, J. (1977). *Problemas de la Mediciòn del Bienestar y Conceptos Afines (Una aplicacion al caso espanol)*. Presidencia del Gobierno, Instituto Nacional de Estadística.

Penco, L., Ivaldi, E., Bruzzi, C., & Musso, E. (2020). Knowledge-based urban environments and entrepreneurship: Inside EU cities. *Cities, 96*, 102443.

Penco, L., Ivaldi, E., & Ciacci, A. (2021). Entrepreneurial ecosystem and well-being in European smart cities: a comparative perspective. *The TQM Journal, 33*(7), 318–350.

Rennes Métropole. (2022). *Rennes Métropole, smart city*. Accessed Jan 23, 2023, from https://metropole.rennes.fr/rennes-metropole-smart-city

Sen, A. (1992). *Inequality reexamined*. Harvard University Press.

Smart Belfast. (2018). *Smart Belfast–Harnessing new technologies and data science in ways that support local economic growth*. Accessed Jan 23, 2023, from https://smartbelfast.city/

Smart Geneva. (2022). *Smart Geneva—L'innovation durable, responsable et citoyenne*. Accessed Jan 23, 2023, from https://www.smart-geneva.ch/en/home

Smart City Graz. (2017). *Smart city Graz 2050*. Accessed Jan 23, 2023, from http://www.smartcitygraz.at/more_vision-fuer-eine-smart-city-graz-2050/

Smart City Oulu. (2020). *Smart city Oulu | Cool city–hot sOulutions*. Accessed Jan 23, 2023, from https://smartcityoulu.com/en/

Smart City Press. (2020, April 21). 8 smallest cities in the world with the soul of A. *Smart city*. Accessed Jan 23, 2023, from https://smartcity.press/small-cities-becoming-smart-cities/

Smart City Wien. (2022). *Smart climate city strategy*. Accessed Jan 23, 2023, from https://smartcity.wien.gv.at/en/strategy/#top

Somarriba, N., & Peña, B. (2009). Synthetic indicators of quality of life in Europe. *Social Indicators Research, 94*(1), 115–133.

Tanguay, G. A., Rajaonson, J., Lefebvre, J. F., & Lanoie, P. (2010). Measuring the sustainability of cities: An analysis of the use of local indicators. *Ecological Indicators, 10*(2), 407–418.

The Global Smart City Knowledge Center. (2018, March 24). *Smart city Antwerp: The European 'Capital of Things'*. Accessed Jan 23, 2023, from https://hub.beesmart.city/city-portraits/smart-city-antwerp

The Global Smart City Knowledge Center (2020, July 27). *Smart city Rotterdam: A leading light in smart innovation*. Accessed Jan 23, 2023, from https://hub.beesmart.city/city-portraits/smart-city-rotterdam-a-leading-light-in-smart-innovation

Townsend, P. (1987). Deprivation. *Journal of Social Policy, 16*, 125–146.

TU Dresden. (2021, June 10). *VAMOS2–Traffic management system VAMOS Dresd*en. Accessed Jan 23, 2023, from https://tu-dresden.de/bu/verkehr/vis/vpa/forschung/Individualverkehr/verkehrsmanagementsystem-vamos-dresden?set_language=en

Wilab. (2022). *Amsterdam–ICT*. ICT. Accessed Jan 23, 2023, from https://www.wilab.org/ICT/smart-cities/amsterdam/

Winkler, P. (1982). Average height in a partially ordered set. *Discrete Mathematics, 39*(3), 337–341.

Wittmann, J., & Brüggemann, R. (2014). A software platform towards a comparison of cars: A case study for handling ratio-based decisions. In R. Brüggemann et al. (Eds.), *Multi-indicator systems and modelling in partial order* (pp. 147–164). Springer.

Zarzosa, P., Sommariba, N. (2013), 'An assessment of social welfare in Spain: Territorial analysis using a synthetic welfare indicator', Social Indicators Research, 111, 1–23, 1.

Zurich City Council. (2018). *Strategy SMART CITY ZURICH*. Accessed Jan 23, 2023, from Smart_City_Zurich_Strategy.pdf

Chapter 5
Policy Implications for Human Well-being

Andrea Ciacci and Enrico Ivaldi

Abstract This chapter aims to take stock of the challenges that policymakers and governance systems have to face in SSC. Specifically, the chapter deals with thorny policy coordination and integration issues. Policy formulation depends on harmonious processes during which multiple stakeholders are involved. Successful policies are contingent on well-balanced supply-side approaches. In addition, a people-centric way of thinking based on new technologies should characterize the policy lines of SSC. To make this effective, policymakers should promote inbound-talent dynamics based on a conducive regulatory environment that overcomes potential barriers to mobility. The chapter ends by providing examples of successful policies that have led to improvements of multiple dimensions in many SSCs worldwide.

Keywords Public policy · Human capital · Well-being

5.1 Policy Actions and Management

Dealing with political and managerial issues within the smart sustainable city (SSC) means thinking about how it would be possible to facilitate the transition to an innovative city model, identifying the distinctive elements that such a city should have in order to effectively define itself as a SSC and remedy any critical issues that may arise during policy development and implementation (Nam & Pardo, 2011).

Governing a SSC implies promoting an approach based on collaboration, cooperation, partnership, citizen engagement, and participation (Coe et al., 2001). Collaboration must involve different functional sectors and parties, from the government to business, academics, non-profit and voluntary organizations, and between jurisdictions within a given geographical area (Anderson & Tregoning, 1998; Lindskog, 2004; Paskaleva, 2009). The United Nations (2018) emphasized the need to empower youth in cities and guide them toward developing entrepreneurship attitudes based on proven business models adapted to their city needs. These business opportunities can result from strategic partnerships and strategies to increase productivity and decrease unemployment. In every case, a role for technologies exists to assist in their achievement.

E. Ivaldi, A. Ciacci, *Smart Sustainable Cities and Knowledge-Based Economy*,
Human Well-Being Research and Policy Making,
https://doi.org/10.1007/978-3-031-25038-5_5

It is evident that SSCs need precise development plans in order for policy development and implementation to succeed. Stakeholders and public actors involved in SSC governance should share concepts (promotional identity and brand), visions, goals, priorities, and strategies (Dirks et al., 2010; Eger, 2009). Complexity and plurality are the elements that characterize many innovative and transformation processes. Therefore, SSC initiatives entail inter-sectoral and inter-institutional collaboration (Albino et al., 2015). Furthermore, the role of the leader is pivotal both to reach the goals included in the process and to strengthen the relations with citizens (Anthopoulos & Fitsilis, 2010; Anthopoulos & Tsoukalas, 2005). However, policy coordination and integration are the main issues that hinder the successful implementation of policies. A city governance system should include and harmonize the stakeholders' potentially diverging visions and needs to reach a better balance.

5.2 Coordination and Integration of Policies

Coordination of policies across spatial scales, organizational practices, and all levels of governance is relevant to develop successful policies and favor innovation in a city (Marceau, 2008). Metropolitan areas are affected by many policies from different levels of government that may be poorly coordinated, fragmented, overlapping, or conflicting. Therefore, controversial outcomes could arise. Integration is not merely for technologies, systems, infrastructure, services, or information but for policies. Packages of policies should prevail over single-focused interventions (Johnson, 2008; Mingardo, 2008).

We can distinguish three policy integration types:

- Sectoral integration relates to the coordination of policy fields and sectors, for example, economic policy, transportation policy, and housing policy (Candel & Biesbroek, 2016).
- Horizontal integration denotes the alignment of policies between actors in an urban area (Paskaleva, 2009). Most metropolitan areas are governed by many municipalities that interact with each other and share resources.
- Vertical integration concerns the coordination between different layers of government, i.e., federal (central or national), state (provincial or regional), local (or municipal), and international context (Candel & Biesbroek, 2016).

The different visions of policy actors at various levels could undermine policy integration. These visions may conflict with each other generating problems or decision-making stalemates (Mingardo, 2008). For example, increasing accessibility to transportation could be detrimental to the urban environment, while air quality improvement might result in restricting accessibility. A challenge for the SSC is to contemporarily maintain economic growth, stay accessible, and improve quality of life. The incompatibility between different visions can depend on the difficulty in achieving a synthesis, compromise, or balance (Nam & Pardo, 2011).

5.3 Demand-Driven or Supply-Side Type of Policy

The successful policies in SSCs are mainly demand-driven or well-balanced between the two approaches rather than supply-driven. Demand-driven policies refer to advanced forms of governance in which the policy development process is concerted among many actors at different levels. On the other side, supply-driven policies indicate a prevailing top-down style of policy development. Supply-side government-driven policies alone are insufficient and need complementing demand-side initiatives. Nam and Pardo (2011, p. 189) write that "policies in successful smart cities are demand-driven rather than supply-driven, or well-balanced between the two approaches. [...] Demand-focused policies may lead to better governance. [...] Policies for a smart city initiative should support collaboration and partnership as a strategy to overcome fragmentation by including key stakeholders." SSC policies must be balanced with more on the demand side and encourage diversity, social networks, and cross-sector innovation. The participation of key stakeholders is decisive in successfully innovating (Hartley, 2005; Greenhalgh et al., 2004, 2005). Demand-side policies are a form able to promote active citizenship and citizen network governance. Citizen participation can happen easily and effectively in a SSC model centered on demand-side policies (Paskaleva, 2009). This approach, sustained by the functionality of the new technologies, gives the government more opportunities to engage the public in a transparent learning environment that provides feedback into governance (Cromer, 2010). Demand-driven policies supported by adequate branding strategies can increase the SSC's appeal to workers and people outside the city. Branding strategies indicate strategic processes (e.g., joint management of physical and virtual identities) that benefit the urban economy (Molinillo et al., 2019, p. 248). The interrelationships between city branding and stakeholder participation are crucial to stimulate investments by local governments (Hospers, 2010; Molinillo et al., 2019). Innovation in the policy dimension requires a branding strategy (Luke et al., 2010). A brand is assimilable to a public promise that a city government makes to urban residents and external people or organizations. Image-making is crucial in facilitating the transition toward an SSC because a more popular brand makes a city well-known outside its bounds (Hospers, 2008). City marketing is necessary for cities that act as a magnet to attract new talent, resources, and investments.

5.4 Policy Integration

Policy integration is a process of "asynchronous and multi-dimensional policy and institutional change within an existing or newly formed governance system that shapes the system's and its subsystems' ability to address a cross-cutting policy problem in a more or less holistic manner" (Candel & Biesbroek, 2016, p. 217). To describe and enhance the policy integration effectiveness, Candel and Biesbroek

(2016) built a policy integration framework around four dimensions, i.e., policy frame, subsystem involvement, policy goals, and policy instruments. The literature states that these dimensions do not move in the same direction since all integration processes differ in development pace. These problems originate due to path dependency that makes dominant subsystems or policy instruments unchanged over time (Pierson, 2000; Streeck & Thelen, 2005). In other words, the logic of consensus prevails over the authenticity of the policy. Another critical issue concerns mutual dependencies and interactions between dimensions. Public policy literature highlights that those policy elements (e.g., type of policy, level of government) and contextual conditions mutually influence in multiple ways (Hall, 1993; Sabatier & Jenkins-Smith, 1993).

The integration policies are based on the *policy frame*, i.e., the definition of the dominant problem of the society perceived by public opinion (Baumgartner & Jones, 2009; Candel & Biesbroek, 2016). It is relevant to consider the policy frame because its absence can pose serious risks. Gieve and Provost (2012), for example, remember how the lack of a coordinated approach between monetary and regulatory policies contributed to the financial crises between 2007 and 2009. Focusing events, policy entrepreneurship, interest mobilizations, and the culture of a governance system are the most common sociopolitical mechanisms that influence the continuity and change of the broader policy frames (Baumgartner & Jones, 2009).

The *subsystem* dimension captures the range of actors and institutions that deal with a policy problem. These actors tend to expand their influence on a broader number of issues. While this tendency would lead to a higher concentration of policies into a sole subsystem, the involvement of different subsystems provides new information, perspectives, and resources that fuel the decision-making process and increase the likelihood of a solution (Candel & Biesbroek, 2016; Jack, 2005).

Another characteristic of decision-making from the subsystem perspective is the power asymmetry during the process. The so-called dominant subsystems (Candel & Biesbroek, 2016) have centrality in the network of interactions and are more influential actors than others (Hartlapp et al., 2012). The dominant subsystems aim to preserve their centrality by increasing their mutual interactions and reducing the interactions with the less engaged subsystems.

Policy integration also depends on intra-governance system substantive and procedural policy instruments. The former allows the allocation of governing resources to affect the distribution, quantities, or types of goods and services provided in society. Procedural instruments indirectly influence outcomes through policy process manipulation (Howlett, 2000) and facilitate the coordination of different subsystems (Jordan & Schout, 2006).

Policy goals represent a pillar of policy integration. They indicate the governance system's adoption of a specific concern to address issues through ad hoc policy and strategy formulation (Howlett & Ramesh, 2003). The level of policy goals integration can vary. An example of low integration of policy goals is the integration of transport policy is hindered by autonomous and sectoral goal-setting by other subsystems (Stead, 2008). On the other hand, the European Commission's coordination system about climate change policies represents the integration of policy goals realized at a high level, which involves climate change mitigation and others in the energy and maritime affairs sectors (Hustedt & Seyfried, 2016). Moreover, policy goals are often perceived differently on temporality or geographical base (Adelle et al., 2009).

A lack of goal consistency can undermine the capacity of a political system to respond to concerns synergistically. An appropriate instrument mix (e.g., financing, policy strategies, legislation, impact assessments, and interdepartmental working groups) can help overcome this inconsistency (Adelle et al., 2009; Feiock, 2013; Karré et al., 2013). The US Community Empowerment regime (1960–1970) represents an example. Its subsystems (e.g., economic development, education, employment, transportation, and welfare) worked in a coordinated way toward urban renewal (Jochim & May, 2010). Another example is the governmental advisory body known as the Finnish Science and Technology Policy Council. This council facilitates policy integration in the science and technology domains (Pelkonen et al., 2008).

5.5 Multilevel People-Centric Approach

Different strategic approaches apply to SSC innovation. Starting from the point that SSC development is related to the city's ability to satisfy the needs of its citizens, the paradigm of people-centric service intelligence emerged overwhelmingly over the last decades (Xu & Geng, 2019). It can be considered a natural shift in the era of big data. The people-centric service intelligence is different from business intelligence. Business intelligence comprises strategies and technologies used by enterprises for data analysis of business information. People-centric service intelligence involves the theories, strategies, and technologies deriving from and complying with the requests and demands of urban citizens. The diffusion of intelligence in SSCs could help to evolve toward civic intelligence (Schuler, 2016). The theoretical framework of people-centric service intelligence can be declined on different levels, as follows:

- Level 0 is highly related to the computational implementation of people-centric service intelligence, for example, knowledge engineering, data mining, artificial intelligence, machine learning, and prediction theories. These theories provide convenient, efficient, and powerful computational capabilities to define, code, embed, and generate service intelligence from people, society, and big data (Xu & Geng, 2019).
- Level 1 is based on the comprehension of human behavior to satisfy the demands and preferences of the local community (Becker & Maiman, 1975; Schwarzer, 2008).
- Level 2 is about the theory of attitude (Katz, 1960; Breckler, 1984). Attitudes redirect the effectiveness and practicability of service intelligence. This approach aims to take action on the fundamental aspects of personal life. The theory of

attitude looks at fully satisfying requests such as safety, health, and economic safety through employment and protection from abnormal events in everyday life.

- Level 3 regards social theory, based on social interactions, networks, and community to maximize the belonging sense in the city (Maslow, 1943).
- In level 4, the set of values is the most relevant factor in influencing the people-centric development of a city (Maslow, 1943). The perception of importance and contribution to other people or society leads to a feeling of esteem, which are crucial dimensions of constructing service intelligence (Xu & Geng, 2019).
- The transition to a people-centric model must be aimed at personal satisfaction (level 5). In this way, the city becomes the place where individuals can meet their expectations. Service intelligence includes the related theory of discovering and learning the culture of people, defining their abilities, potential, and aspirations, while reaching equality in surrounding relationships and society (Xu & Geng, 2019).

5.6 Managerial Innovativeness for Public Policy Innovation

Therefore, in the previous section, we briefly described the relevance of new technologies in the SSC development process. To speed up this development, SSC governments should integrate ICT use as an enabler factor of their traditional modus operandi. According to Moon and Norris (2005), managerial innovativeness is the most compelling reason to explain the adoption by municipal governments of new ICTs in their core functions. Managerial innovation affects technological and administrative innovation (Walker et al., 2011). Successful organizational change in the public sector should be managed (Fernández & Rainey, 2006). A SSC is the application of intelligence to city management (Nam & Pardo, 2011).

Both vertical and horizontal governance for entrepreneurship appears to present challenges for business actors, policymakers, and researchers. In order to adopt effective policy measures, it is necessary to bear in mind the uncontrollable factors that, starting from the context in which the actors operate, make the outcomes of the processes unpredictable. For this reason, it is not a proper choice to study the potential effects of political actions applied in other countries, cities, and the environment. To underline this latent uncertainty factor, Lerner (2014, p. 1) states that the "public sector's pursuit of entrepreneurial growth is a massive casino where bets are made with few guarantees of good returns."

Mainly there are integration problems: innovation means increasing the level of complexity, so social adaptation to new, complex, and rapidly evolving realities toward progress could be problematic (Caragliu et al., 2009; Hollands, 2008; Jennings, 2010). These processes can originate due to persistent failures in the public sector, policy (policy integration, branding for marketing, demand-focused initiative), management, and organization (enterprise architecture, cross-organizational management, role of leadership). They are mostly related to poor risk management capacity. In addition, due to accountability processes, there is a heightened aversion to risk on the part of public administrations and, more generally, the political classes, which prefer to adopt short-term measures rather than long-term projects in order not to lose consensus (Potts & Kastelle, 2010; Cromer, 2010).

5.7 Regulatory Barriers Contrasting Inbound-Talent Dynamics

Smart cities must leverage their attractive factors to generate centripetal thrusts that lead to inbound dynamics of skilled workers. From this perspective, over technological primacy, human capital represents another differentiation lever to prevail over counterparts.

Many recent perspectives on the future scenarios of smart cities remark on the importance of human capital (Angelidou, 2015). For instance, the mainstream views consider human-centered and technology-people combined scenarios as prevailing ones. Following these conceptualizations, the city's competitiveness lies in its ability to attract skilled workers and deploy their capabilities together with advanced technology to co-construct the urban evolutionary paradigm (Andreani et al., 2019).

The phenomenon based on the competition between cities to attract highly skilled human and knowledge capital is known as talent competition. The literature recognized that not only talent competition refers to firms in their industry (Delgado-Verde et al., 2016; Siepel et al., 2017) but also to cities in the worldwide scenario. Today, the talent competition is global and multimarket and looks at multiple, diversified knowledge and skills. Attracting highly skilled human capital is a critical resource for the urban sustainable competitive advantage (Hajek et al., 2022; Thite, 2011).

In particular, SSCs built on knowledge-based economies are more prone to search for highly educated professionals (e.g., physicians, engineers, architects, experts in digital transformation and data analytics, etc.) to maximize the potential of their service sectors, creativity, and entrepreneurial fabrics. Therefore, cities should incentivize such inbound-talent dynamics (i.e., inward mobility) to reach the competition frontier.

However, in some cases, multiple barriers may hinder the individual propensity to move toward other countries and cities. In some cases, identifying and applying suitable enabling mechanisms may be difficult because they depend on the simultaneous coordination (or multilevel regulatory interventions) enacted by different cities worldwide. These limitations jeopardize the possibility of stimulating inward mobility mechanisms, reducing the city's reservoir of human and cognitive capital.

In this sense, one of the most challenging obstacles concerns professional qualification recognition. In their study on the European labor market, Krause et al. (2017) identified professional qualification recognition as one of the main barriers to a well-developed model of inter-city mobility. Put differently, highly skilled workers moving from one city to another may encounter difficulties in receiving recognition for their acquired education and training qualifications. This is particularly true for occupations implying high levels of specialized knowledge (e.g., physicians, engineers, data scientists, architects, etc.) since the educational and training requirements vary from one country to another and from one city to another. These circumstances compel the specialized workers to take additional exams to obtain the necessary credentials. The negative effect of such regulatory limitations is to discourage inter-city mobility or impede a professional from practicing the occupation in a new city.

To address this issue, in 2008, the European institutions, driven by pressure from multinational firms, adopted a recommendation leading the member states to provide information on their national qualification systems and qualification certificates. This initiative aimed to establish a meta-framework linking misaligned national education systems by increasing qualification transparency.

Such an exemplary case shows the relevance of private actors in promoting political initiatives through bottom-up approaches. Overall, these policies are grounded on economic rationales aimed at emphasizing human capital and involve education and political reforms (Kleibrink, 2011).

Another measure to overcome misalignments at a system level may consist of mutual recognition agreements between two or more countries through which professional credentials are recognized without completing additional training courses. Differently, policymakers may decide to make the foreign credentials recognition contingent upon the completion of further assessments. In addition, issuing temporary visas can stimulate the propensity to transfer to practice their occupation in a new city.

Other factors contrasting inter-city mobility are related to the lack of transparency in job offerings as well as the difficulties in matching job demand and offering (Krause et al., 2017). In this regard, the private actors may take the initiative to create occupational communities, i.e., social structures in which people support each other in planning their career paths by sharing information and experiences (Curşeu et al., 2021). At the same time, policymakers should promote job offering diffusion and transparency and advertisement systems to increase their pervasiveness.

Talent retaining is another effective strategy to win the talent competition. In this way, SSC may invest in retaining their talents after graduation. Such a strategy should aim to valorize the careers and define connection channels between public and university institutions and private sectors. Finally, as argued in the previous chapters of this book, policymakers should put livability-related issues first in their priority agenda to incentivize talent attraction.

More in general, to win the competition for talent, policymakers should put in place superior administrative processes by improving migration strategies (BCG Henderson Institute, 2022). In a nutshell, policymakers must timely adapt their regulatory environment to unleash the actual potential of their cities. In doing this, they should adopt policies to make their city a catalyst for a skilled workforce, creating a conducive environment for key stakeholders (Kummitha & Crutzen, 2019).

5.8 SSC Policies Worldwide

This section focuses on practical examples of policies based on global cities. In Table 5.1, we provide some examples of policies concerning the SSC dimensions analyzed in chap. 4 of this book, i.e., ICT and transport, people, sustainability, and government. We analyze the purpose of the policies, expected outcomes, and related examples of success worldwide for each dimension.

SSCs invest many resources in developing advanced ICT systems. First, ICT facilitates the strategic positioning of a SSC in the global scenario (TWI, 2022). ICT plays an enabler role in the SSC design. For instance, ICT facilitates the creation of more innovative entrepreneurial ecosystems. ICT pervasively impacts daily life (Bibri & Krogstie, 2017). Sensor systems installed on traffic lights, for example, facilitate the management of traffic flows. Importantly, ICT enables higher standards of connectivity in a city. In parallel, investments in the modernization of the public transport system promote spatial connectivity. More highly connected cities are prominent in the interregional networks (Verginer & Riccaboni, 2021). Overall, the purpose of ICT and transport policies is to create digitized environments to increase more connected cities in terms of Internet and spatial connections.

Sustainability is a crucial dimension of a SSC. Making a SSC more sustainable means creating sustainable energy infrastructure to unleash the potential of renewables, finding solutions against climate change, reducing carbon emissions, and embracing a circular economy model at the city level (European Commission, 2022). Singapore, Washington DC, Hong Kong, and Vancouver are paradigmatic examples of cities highly committed to creating sustainable and circular models (Blaise, 2022; Hamza, 2021; Igini, 2021).

SSC policies have a central role in improving the citizens' quality of life, ensuring jobs for all citizens, and, more in general, addressing the future needs of the population (European Commission, 2022). These policies aim to find an intergenerational balance by adopting measures for young and aged people. For instance, policymakers adopt measures to attract young human capital as a skilled labor force for a knowledge-based economy and improve healthcare through IT potential. These policies contribute to creating a livable and accessible city for all. SSCs make efforts to protect the privacy and cybersecurity of citizens' data. In fact, the steady interconnectivity characterizing SSCs exposes citizens to risks associated with the breach of privacy and data theft (Bibri, 2018).

Concerning the government dimension, the main purpose of the policies is to unite different cities and stakeholders into a strong network (European Commission, 2022). Therefore, coordination and integration are at the heart of government policy purposes. Specifically, these policies aim to create public-public partnerships to align different needs and resources. In addition, strong networks of stakeholders imply full participation by citizens in political processes. Therefore, these political processes are more similar to participative and inclusive governance rather than top-down processes. It leads to bottom-up approaches that put citizens at the center of the decision-making process stage. ICT-powered data accessibility makes citizens

Table 5.1 SSC policies and successful examples

Dimension	Policy purposes	Expected outcomes	World successful examples	Sources
ICT and transport	• Promoting innovation • Positioning itself as a SSC • Reducing urban traffic • Geolocation of public facilities • Modernization of public transport • Increasing fiber network coverage	These policies aim to invest in digitization to create facilities for all citizens. Special attention is devoted to fast connections and effective and efficient public transportation services. In fact, cities can reduce traffic through transportation network improvement and reduction in transportation by private means. It would benefit the city's economy and employment. Finally, in the implementation of digital strategies, the role of the city's different specificities takes on a key role: The goal is to listen to each individual specificity to create a common and functional network	• The city-state of Singapore deploys IoT cameras to monitor the cleanliness of public spaces, population density, and the movement of cars. These policies make Singapore one of the leaders in the race to develop SSCs. Singapore also includes real-time monitoring systems for water consumption, waste management, and electricity use. In order to safeguard the health and Well-being of older adults, there are also monitoring systems and autonomous car testing. • San Diego has deployed 3200 sensors to improve environmental awareness, public safety, parking, and traffic flow. Solar-to-electric charging stations help electric vehicles, and networked cameras keep an eye out for traffic issues and criminal activity.	Zurich City Council (2018); Rennes Métropole (2022); Cardiff Council's Smart City Road Map (2021); City of Stockholm (2022); TWI (2022)
People	• Improving the citizens' quality of life • Ensuring jobs	The goal of policies in this dimension is to create a livable and	• Dubai offers smart buildings, utilities, education, tourist	European Commission (2022); Zurich City Council (2018);

(continued)

Table 5.1 (continued)

Dimension	Policy purposes	Expected outcomes	World successful examples	Sources
	for all citizens • Meeting the future needs of the population • Increasing culture and sociability • Youth policies • Improving the privacy and cybersecurity of citizens' personal data	accessible city for all. For this reason, much space is given to youth and elderly policies. In addition, culture and the creation of strong social network ties are the foundations for attracting social capital, social capital that will have as repercussions an increase in city competitiveness and an increase in the economy and jobs	alternatives, healthcare solutions (e.g., telemedicine), and traffic monitoring systems. • Australia's New South Wales implemented a smart pilot program to enhance urban inhabitants' access to healthcare. Residents in what the NSW government refers to as "integrated healthcare" belong to a $180 million project developed to restructure care. Residents are aging and placing additional strain on the present system. The initiative includes funding for developing technologies that better track patients and care for monitoring purposes, promotion of HEALTHeNet, the connection of various healthcare providers and patient data across the region, and the shared care collaboration scheme, which encourages families to collaborate with professionals.	Smart City Press (2020); Smart City Wien (2022); Smart Belfast (2018); TWI (2022); Osborne (2016)
Sustainability	• Creation of increasingly	The goal of policies in this	• Singapore has demonstrated over	European Commission (2022);

(continued)

Table 5.1 (continued)

Dimension	Policy purposes	Expected outcomes	World successful examples	Sources
	sustainable energy infrastructure • Transformation of the energy system • Finding solutions to climate change • Reducing carbon emissions • Implementation of circular economy • Making the entire city grounds energy self-sufficient through renewable energy	dimension is to reduce the impact on the environment and climate change. For this reason, the basis of these policies is the establishment of a circular economy model, the encouragement of the use of energy from renewable sources, and the creation of a self-sufficient city energy model	the past ten years that it has the ability to lead the world in sustainability. It is one of the most virtuous examples of green areas, smartness, and sustainability at the global level. In order to achieve these results, Singapore adopted technology innovations, environmentally friendly initiatives, and smart policies across all major sectors. • The fundamental thrust of Washington D.C.'s smart innovation programs has been its commitment to finding solutions to climate change and promoting environmental sustainability. These initiatives actually changed something. The district has decreased its overall carbon impact by 31% since 2006. • In 2015, when buildings accounted for 90% of the city's power use, Hong Kong's major projects got underway. Hong Kong has so far been able to reduce the overall	Making City (2019); Smart City Wien (2022); Cardiff Council's Smart City Road Map (2021); Malmö stad (2021); Igini (2021); Hamza (2021); Blaise (2022)

(continued)

Table 5.1 (continued)

Dimension	Policy purposes	Expected outcomes	World successful examples	Sources
			carbon footprint by 35%, recover 1.64 million tons of municipal solid waste, and decrease the electricity consumption of government buildings by 7.8% through smart building technologies, waste management, sensors to monitor pollution, and educational campaigns. • Vancouver, a significant metropolis in North America, uses more than 90% sustainable energy, making it the largest city in the area to generate the fewest greenhouse emissions per resident. Vancouver has prospered thanks to real-time data, technology, and improvements in connectivity, sustainability, and ease.	
Government	• Unite different cities and stakeholders into a strong network • Increasing the competitiveness of European cities and industries • Creation of public-private partnerships	The goal of policies in this dimension is to put citizens at the center of decision-making processes and to create a robust public-private partnership model. Therefore, these policies aim to create bottom-	• Online service management is available to public housing tenants via the MyNYCHA mobile app and web interface. It covers more than 300 public construction projects in new York City.	European Commission (2022); Etienne (2019); Thales Group (2022); We Build Value (2020)

(continued)

Table 5.1 (continued)

Dimension	Policy purposes	Expected outcomes	World successful examples	Sources
	• Close collaboration between politics, business, science, culture, and society • Ensuring fair transparency and accountability • Full citizens' participation in political processes • Data accessibility	up processes through which citizens inspire the policies and the government implements the decisions. The creation of such a model combined with close collaboration between different public and private stakeholders would increase the competitiveness of cities and strengthen the economy and coordination between European SSCs	MyNYCHA, a free program introduced in 2015, gives residents control over the restoration process. Online work tickets can be submitted, scheduled, and managed by residents. They may pay their rent, examine inspection schedules, and sign up for notifications for issues in their projects. • To build a list of short- and medium-term initiatives targeted at smartly growing the city, Toronto has established a think tank for huge organizations like IBM together with start-up incubators.	

aware of the city and resident needs. Data represents a relevant source of information to overcome informative asymmetries between the different stakeholders.

5.9 Conclusions

This chapter provides insights into the processes of policy development and implementation. It describes the relevance of policy coordination and integration to find the most valuable harmonization between multiple levels of involved actors. This chapter also highlights the potential risks that policymakers and political systems fall into during policy development when coordination and integration fail. In summary, policymakers must devote attention to the process side of policy development and implementation to obtain the desired outcomes. In addition, from this chapter, the necessity to open the boundaries of policy development to stakeholders at different levels emerges. Promoting participative policies based on shared purposes appears the best way to align the stakeholders' intentions. In addition, reaching the right balance between top-down (supply-side) and bottom-up (demand-side) policies and defining the governance boundaries appear as two prominent issues to ensure the effectiveness of the public policy formulation. The chapter ends with a multidimensional description of exemplary urban cases. The main implications arising from these paragraphs refer to the necessity for an SSC to undertake an evolution along multiple dimensions. In fact, the transition toward a model of SSC requires progressive and diffused transformations. In this regard, the development of a dimension (e.g., ICT) is functional to the development of the other dimensions. This evidence raises the issue of ICT ubiquity (Suopajärvi, 2015).

Dubai
Country: United Arab Emirates
Inhabitants: 3,515,813
Density: 2200/km^2 (5700/sq. mi)
Total area: 1610 km^2 (620 sq. mi)

Within the next 40 years, one of the largest cities of the United Arab Emirates (UAE) wants to transform its existing metropolis, which is already experiencing tremendous economic and social development, into one of the happiest cities on earth. In order to transform Dubai into a truly Smart City in the near future, numerous measures have been implemented.

Aiming to improve the quality of life for workers and citizens, the municipal administration made the city's government offices completely paperless in 2021. This outcome was attained through the total digitalization of all documents.

In terms of innovation, emphasis is put on blockchain technology, which has the potential to transform the web's future and improve the speed and efficiency of transactions and connections. Government efficiency, industrial production, and global leadership are used to meet the goals set by the city's strategy. The UAE as a whole and Dubai have an ambitious medium-long-term goal of becoming one of the most well-known start-up incubators in the world. As aforementioned, in addition to being a smart city, Dubai also wants to become the happiest city in the world. To that end, a "happiness agenda" has been created, which outlines the policies the city plans to implement to enhance the quality of life for its citizens and the millions of tourists who visit it annually.

A unique "happiness index" designed to capture the requirements of people interviewed and convert the information gathered into policy will be used to gauge the level of happiness of both workers and tourists.

(continued)

However, artificial intelligence (AI), a crucial component of Dubai's sustainable strategy, still presents ethical issues that need to be resolved. For this reason, the city has built an "ethical AI toolkit" to support relevant sectors and an AI research center to test solutions for integrating the machine learning tool into governmental, industrial, and tourist services.

The UAE pass, a digital national identity card that gives people access to all government services and the ability to digitally sign in for the purchase of a car or a property via an app, is the first step to digitalizing municipal services. Along with this ground-breaking app, the local government has also created a smart bot called "Rashid" with artificial intelligence that can assist individuals in their involvement in bureaucratic processes and respond to their inquiries. The AI lab also created the Dubai pulse and DubaiNow portal: The former is a data collection platform capable of providing information in open access to citizens and businesses who request it. The portal instead refers to a single platform capable of guaranteeing access to 55 different government services via a simple click.

Dubai has started the process of becoming a real smart city of the future, and it relies on blockchain technology, AI, and open data to accomplish its main goals.

References

Adelle, C., Pallemaerts, M., & Chiavari, J. (2009). *Climate change and energy security in Europe: Policy integration and its limits.* Swedish Institute for European Policy Studies.

Albino, V., Berardi, U., & Dangelico, R. M. (2015). Smart cities: Definitions, dimensions, performance, and initiatives. *Journal of Urban Technology, 22*, 3–21.

Anderson, G., & Tregoning, H. (1998). Smart growth in our future? In Urban Land Institute (Ed.), *ULI on the future: Smart growth* (pp. 4–11). Urban Land Institute.

Andreani, S., Kalchschmidt, M., Pinto, R., & Sayegh, A. (2019). Reframing technologically enhanced urban scenarios: A design research model towards human centered smart cities. *Technological Forecasting and Social Change, 142*, 15–25.

Angelidou, M. (2015). Smart cities: A conjuncture of four forces. *Cities, 47*, 95–106.

Anthopoulos, L., & Fitsilis, P. (2010). From online to ubiquitous cities: The technical transformation of virtual communities. In A. B. Sideridis & C. Z. Patrikakis (Eds.), *Next generation society: Technological and legal issues (Proceedings of the Third International Conference, eDemocracy 2009)* (Vol. 26, pp. 360–372). Springer. Accessed Jan 23, 2023, from http://www.springerlink.com/content/g644776482968k36/fulltext.pdf

Anthopoulos, L., & Tsoukalas, I. A. (2005). The implementation model of a digital city. *Journal of EGovernment, 2*(2), 91–110.

Baumgartner, F. R., & Jones, B. D. (2009). *Agendas and instability in American politics*. University of Chicago Press.

BCG Henderson Institute. (2022). *How cities can win the competition for talent—and residents.* Accessed Jan 23, 2023, from https://bcghendersoninstitute.com/how-cities-can-win-the-competition-for-talent-and-residents/

Becker, M. H., & Maiman, L. A. (1975). Sociobehavioral determinants of compliance with health and medical care recommendations. *Medical Care, 134*, 10–24.

Bibri, S. E. (2018). The IoT for smart sustainable cities of the future: An analytical framework for sensor-based big data applications for environmental sustainability. *Sustainable Cities and Society, 38*, 230–253.

Bibri, S. E., & Krogstie, J. (2017). Smart sustainable cities of the future: An extensive interdisciplinary literature review. *Sustainable Cities and Society, 31*, 183–212.

Blaise, H. (2022, 22 April). *Top 10: Smart cities driving sustainability around the world.* Home | Sustainability Magazine. Accessed Jan 23, 2023, from https://sustainabilitymag.com/top10/top-10-smart-cities-around-the-world

Breckler, S. J. (1984). Empirical validation of affect, behavior, and cognition as distinct components of attitude. *Journal of Personality and Social Psychology, 47*, 1191–1205.

Candel, J. J. L., & Biesbroek, R. (2016). Toward a processual understanding of policy integration. In *Policy Sciences* (Vol. 49, pp. 211–231). Springer.

Caragliu, A., Bo, C. D., & Nijkamp, P. (2009). *Smart cities in Europe*. In 3rd Central European Conference in Regional Science, pp. 45-60.

Cardiff Council's Smart City Road Map. (2021). *Smart Cardiff.* Available on: 0000765 RC Smart Cities 2019 FIN craig.indd (smartcardiff.co.uk).

City of Stockholm. (2022). *The smart city–city of Stockholm*. Start–City of Stockholm. Accessed Jan 23, 2023, from https://international.stockholm.se/city-development/the-smart-city/

Coe, A., Paquet, G., & Roy, J. (2001). E-governance and smart communities: A social learning challenge. *Social Science Computer Review, 19*(1), 80–93.

Cromer, C. (2010). Understanding web 2.0's influences on public e-services: A protection motivation perspective. *Innovation: Management, Policy & Practice, 12*(2), 192–205.

Curşeu, P. L., Semeijn, J. H., & Nikolova, I. (2021). Career challenges in smart cities: A sociotechnical systems view on sustainable careers. *Human Relations, 74*(5), 656–677.

Delgado-Verde, M., Martín-De Castro, G., & Amores-Salvadó, J. (2016). Intellectual capital and radical innovation: Exploring the quadratic effects in technology-based manufacturing firms. *Technovation, 54*, 35–47.

Dirks, S., Gurdgiev, C., & Keeling, M. (2010). *Smarter cities for smarter growth: How cities can optimize their systems for the talent-based economy*. IBM Global Business Services. Accessed Jan 23, 2023, from ftp://public.dhe.ibm.com/common/ssi/ecm/en/gbe03348usen/GBE03348USEN.PDF

Eger, J. M. (2009). Smart growth, smart cities, and the crisis at the pump a worldwide phenomenon. *The Journal of E-Government Policy and Regulation, 32*(1), 47–53.

Etienne S. (2019, September 27). *Luxembourg-Ville, L'exemple Même De La Smart City*. smartcitiesmag.lu. Accessed Jan 23, 2023, from https://smartcitiesmag.lu/web/luxembourg-ville-lexemple-meme-de-la-smart-city/

European Commission. (2022). *Smart cities. European Commission–European Commission*. Accessed Jan 23, 2023, from https://ec.europa.eu/info/eu-regional-and-urban-development/topics/cities-and-urban-development/city-initiatives/smart-cities_en

Feiock, R. C. (2013). The institutional collective action framework. *Policy Studies Journal, 41*(3), 397–425.

Fernández, L. M., & Rainey, H. G. (2006). Theory to practice. *Public Administration Review, 66*(2), 1–25.

Gieve, J., & Provost, C. (2012). Ideas and coordination in policymaking: The financial crisis of 2007–2009. *Governance, 25*(1), 61–77.

Greenhalgh, T., Robert, G., Macfarlane, F., Bate, P., Kyriakidou, O., & Peacock, R. (2005). Storylines of research in diffusion of innovation: A meta-narrative approach to systematic review. *Social Science & Medicine, 61*(2), 417–430.

Greenhalgh, T., Robert, G., Macfarlane, F., Bate, P., & Kyriakidou, O. (2004). Diffusion of innovations in service organizations: Systematic review and recommendations. *Milbank Quarterly, 82*(4), 581–629.

Hajek, P., Youssef, A., & Hajkova, V. (2022). Recent developments in smart city assessment: A bibliometric and content analysis-based literature review. *Cities, 103709*, 103709.

Hall, P. A. (1993). Policy paradigms, social learning, and the state: The case of economic policymaking in Britain. *Comparative Politics, 25*(3), 275–296.

Hamza, M. (2021, 18 October). *These are the top 20 sustainable smart cities in the world*. Sensors & IoT infrastructure | disruptive technologies. Accessed Jan 23, 2023, from https://www.disruptive-technologies.com/blog/the-top-20-sustainable-smart-cities-in-the-world

Hartlapp, M., Metz, J., & Rauh, C. (2012). Linking agenda setting to coordination structures: Bureaucratic politics inside the European Commission. *Journal of European Integration, 35*(4), 425–441.

Hartley, J. (2005). Innovation in governance and public services: Past and present. *Public Money & Management, 25*(1), 27–34.

Hollands, R. G. (2008). Will the real smart city please stand up? *City, 12*(3), 303–320.

Hospers, G.-J. (2008). Governance in innovative cities and the importance of branding. *Innovation: Management, Policy & Practice, 10*(2–3), 224–234.

Hospers, G. J. (2010). Making sense of place: From cold to warm city marketing. *Journal of Place Management and Development, 3*(3), 182–193.

Howlett, M. (2000). Managing the "hollow state": Procedural policy instruments and modern governance. *Canadian Public Administration, 43*(4), 412–431.

Howlett, M., & Ramesh, M. (2003). *Studying public policy: Policy cycles and policy subsystems*. Oxford University Press.

Hustedt, T., & Seyfried, M. (2016). Co-ordination across internal organizational boundaries: How the EU Commission coordinates climate policies. *Journal of European Public Policy, 23*(6), 888–905.

Karré, P. M., van der Steen, M., & van Twist, M. (2013). Joined-up government in The Netherlands: Experiences with program ministries. *International Journal of Public Administration, 36*(1), 63–73.

Katz, D. (1960). The functional approach to the study of attitudes. *Public Opinion Quarterly, 24*, 163–204.

Kleibrink, A. (2011). The EU as a norm entrepreneur: The case of lifelong learning. *European Journal of Education, 46*(1), 70–84.

Krause, A., Rinne, U., & Zimmermann, K. F. (2017). European labor market integration: What the experts think. *International Journal of Manpower, 38*(7), 954–974.

Kummitha, R. K. R., & Crutzen, N. (2019). Smart cities and the citizen-driven internet of things: A qualitative inquiry into an emerging smart city. *Technological Forecasting and Social Change, 140*, 44–53.

Igini, M. (2021, 26 August). *How sustainable cities like Singapore succeed in green urban development*. Earth.Org. Accessed Jan 23, 2023, from https://earth.org/how-sustainable-cities-like-singapore-succeed-in-green-urban-development/

Jack, S. L. (2005). The role, use and activation of strong and weak network ties: A qualitative analysis. *Journal of Management Studies, 42*(6), 1233–1259.

Jennings, P. (2010). Managing the risks of smarter planet solutions. *IBM Journal of Research and Development, 54*(4), 1.

Jochim, A. E., & May, P. J. (2010). Beyond subsystems: Policy regimes and governance. *Policy Studies Journal, 38*(2), 303–327.

Johnson, B. (2008). Cities, systems of innovation and economic development. *Innovation: Management, Policy & Practice, 10*(2–3), 146–155.

Jordan, A., & Schout, A. (2006). *The coordination of the European Union: Exploring the capacities of networked governance*. Oxford University Press.

Lerner, J. (2014). *Entrepreneurship, public policy, and cities*. World Bank Policy Research Working Paper, (6880).

Lindskog, H. (2004). *Smart communities initiatives*. In Proceedings of the 3rd ISOneWorld Conference (Las Vegas, NV, Apr 14-16). Accessed Jan 23, 2023, from http://www.heldag.com/articles/Smart%20communities%20april%202004.pdf

Luke, B., Verreynne, M., & Kearins, K. (2010). Innovative and entrepreneurial activity in the public sectòr: The changing face of public sector institutions. *Innovation: Management, Policy & Practice, 12*(2), 138–153.

Malmö stad. (2021, 31 March). *Theme sustainable city*. Startsidan–Malmö stad. Accessed Jan 23, 2023, from https://malmo.se/Welcome-to-Malmo/Technical-visits/Theme-Sustainable-City.html

Making City. (2019). *GRONINGEN–making city*. Accessed Jan 23, 2023, from https://Makingcity.Eu/Groningen/. https://makingcity.eu/groningen/

Marceau, J. (2008). Introduction: Innovation in the city and innovative cities. *Innovation: Management, Policy & Practice, 10*(2–3), 136–145.

Maslow, A. H. (1943). A theory of human motivation. *Psychological Review, 50*, 370–396.

Mingardo, G. (2008). Cities and innovative urban transport policies. *Innovation: Management, Policy & Practice, 10*(2–3), 269–281.

Molinillo, S., Anaya-Sánchez, R., Morrison, A. M., & Coca-Stefaniak, J. A. (2019). Smart city communication via social media: Analysing residents' and visitors' engagement. *Cities, 94*, 247–255.

Moon, M. J., & Norris, D. F. (2005). Does managerial orientation matter? The adoption of reinventing government and e-government at the municipal level. *Information Systems Journal, 15*(1), 43–60.

Nam, T., & Pardo, T. A. (2011). *Conceptualizing smart city with dimensions of technology, people, and institutions*. In 12th Annual International Digital Government Research Conference: Digital government innovation in challenging times, pp. 282-291.

Osborne, C. (2016, 19 October). The top 10 smart city citizen projects around the world. *ZDNET*. Accessed Jan 23, 2023, from https://www.zdnet.com/pictures/the-top-10-smart-city-citizen-projects-around-the-world/6/

Paskaleva, K. A. (2009). Enabling the smart city: The progress of city e-governance in Europe. *International Journal of Innovation and Regional Development, 1*(4), 405–422.

Pelkonen, A., Teräväinen, T., & Waltari, S.-T. (2008). Assessing policy coordination capacity: Higher education, science, and technology policies in Finland. *Science and Public Policy, 35*(4), 241–252.

Pierson, P. (2000). Increasing returns, path dependence, and the study of politics. *American Political Science Review, 94*(2), 251–267.

Potts, J., & Kastelle, T. (2010). Public sector innovation research: What's next? Innovation: Management. *Policy & Practice, 12*(2), 122–137.

Rennes Métropole. (2022). *Rennes Métropole, Smart City*. Accessed Jan 23, 2023, from https://metropole.rennes.fr/rennes-metropole-smart-city

Sabatier, P. A., & Jenkins-Smith, H. S. (1993). *Policy change and learning: An advocacy coalition approach*. Westview Press.

Schuler, D. (2016). Smart cities+smart citizens= civic intelligence? In G. Concilio & F. Rizzo (Eds.), *Human smart cities*. Springer International Publishing.

Schwarzer, R. (2008). Modeling health behavior change: How to predict and modify the adoption and maintenance of health behaviors. *Applied Psychology, 57*, 1–29.

Siepel, J., Cowling, M., & Coad, A. (2017). Non-founder human capital and the long-run growth and survival of high-tech ventures. *Technovation, 59*, 34–43.

Smart Belfast. (2018). *Smart Belfast–Harnessing new technologies and data science in ways that support local economic growth*. Smart Belfast–Harnessing new technologies and data science in ways that support local economic growth. Accessed Jan 23, 2023, from https://smartbelfast.city/

Smart City Press. (2020, April 21). *8 smallest cities in the world with the soul of a*. Smart City. Accessed Jan 23, 2023, from https://smartcity.press/small-cities-becoming-smart-cities/

Smart City Wien. (2022). *Smart climate city strategy*. Accessed Jan 23, 2023, from https://smartcity.wien.gv.at/en/strategy/#top

Stead, D. (2008). Institutional aspects of integrating transport, environment and health policies. *Transport Policy, 15*(3), 139–148.

Streeck, W., & Thelen, K. (2005). *Beyond continuity: Institutional change in advanced political economies*. Oxford University Press.
Suopajärvi, T. (2015). Past experiences, current practices and future design: Ethnographic study of aging adults' everyday ICT practices—And how it could benefit public ubiquitous computing design. *Technological Forecasting and Social Change, 93*, 112–123.
Thales Group. (2022). *Secure, sustainable smart cities and the IoT*. Accessed Jan 23, 2023, from https://www.thalesgroup.com/en/markets/digital-identity-and-security/iot/inspired/smart-cities
Thite, M. (2011). Smart cities: Implications of urban planning for human resource development. *Human Resource Development International, 14*(5), 623–631.
TWI. (2022). *What is a smart city?–definition and examples*. Joining Innovation with Expertise–TWI. Accessed Jan 23, 2023, from https://www.twi-global.com/technical-knowledge/faqs/what-is-a-smart-city#Examples
United Nations. (2018). *DG 11 synthesis report 2018: Tracking progress towards inclusive, safe, resilient and sustainable cities and human settlements*.
Verginer, L., & Riccaboni, M. (2021). Talent goes to global cities: The world network of scientists' mobility. *Research Policy, 50*(1), 104127.
Walker, R. M., Damanpour, F., & Devece, C. A. (2011). Management innovation and organizational performance: The mediating effect of performance management. *Journal of Public Administration Research and Theory, 21*(2), 367–386.
We Build Value. (2020, 3 June). *Smart cities in the world: Examples and ranking–we build value*. Accessed Jan 23, 2023, from https://www.webuildvalue.com/en/megatrends/smart-city-world.html
Xu, H., & Geng, X. (2019). People-centric service intelligence for smart cities. *Smart Cities, 2*, 135–152. https://doi.org/10.3390/smartcities2020010
Zurich City Council. (2018). *Strategy SMART CITY ZURICH*. Accessed Jan 23, 2023, from Smart_City_Zurich_Strategy.pdf

Chapter 6
The Cities of the Future

Andrea Ciacci and Enrico Ivaldi

Abstract This chapter focuses on the theoretical strand that defines future cities according to the meanings of "cognition" and "sustainability." In particular, the developments of cities toward cognitive and sustainable evolution are identified, from their origins to their evolutionary trajectory. Each new city paradigm ("digital," "smart," "sustainable," "resilient," "15 minutes," and "circular") is treated individually in relation to its stage of evolution. The evolutionary steps of the city of the future all depend on a single key element: technology. In fact, the city can take on different forms and evolve over time based on the degree of use of technological innovations and their penetration into the social, economic, and environmental fabric.

Keywords Digital city · Resilient city · Circular city · 15-minute city

6.1 The Increasing Urbanization

To understand the future of cities, it is relevant to start from their origins, defining the basic concept and briefly describing the evolution up to the present day. Nowadays there is no univocal definition of a city, and it varies from nation to nation according to the criteria selected: Maunier (1910) tried to categorize the existing definitions of cities and, through a work of synthesis, to provide his own. His study, published in the American Journal of Sociology, found that the definitions of cities existing at the time were essentially divided into two groups: those based on special characteristics and those based on multiplicity characteristics.

In the first group, cities are defined based on characteristics related to morphology (extension of the city, number of inhabitants, etc.), demography (birth or marriage rate), jurisdiction (municipal and commercial law), and functionality (presence of a certain number of activities). The second group includes the definitions that distinguish cities based on the simultaneous presence of a multiplicity of characteristics of the same nature: for example, a city that presents at the same time two or more characteristics belonging to the first group.

E. Ivaldi, A. Ciacci, *Smart Sustainable Cities and Knowledge-Based Economy*,
Human Well-Being Research and Policy Making,
https://doi.org/10.1007/978-3-031-25038-5_6

The definition of city given by Maunier's work is therefore the following:

> a complex community of which the geographic localization is especially limited in relation to the city's size (volume), of which the amount of territory is relatively small with reference to the number of human beings. (Maunier, 1910).

The criterion for defining cities based on the number of inhabitants (morphological characteristics) was also analyzed by Dijkstra et al. (2018), who reported how it varies from country to country. 100 out of 233 countries examined in the study identify a minimum population threshold to define cities: 85% of them use a threshold of 5000 inhabitants or lower (such as India with 2000). Few countries, on the other hand, have more restrictive selection criteria for the definition of a city: Mali has a threshold of 30,000 inhabitants, while in the Far East, Japan and China lead this ranking (threshold of 50,000 and 100,000 inhabitants, respectively).

As for the historical evolution of the cities, it should be reported that the first forms of human settlement recognized date back to the period of the so-called Neolithic revolution, dating back to about 10,000 years ago. In this historical context, the foundations were created for the birth of future cities: following the development of agriculture, there was in fact a significant increase in food production, the food surplus produced allowed the first forms of exchange between people, and this attracted more and more individuals resulting in a greater population density. These ancestral settlements then evolved, over time, into inhabited centers of increasing complexity: the population was always greater in rural areas than in cities until 2007.

The industrial revolution in the nineteenth century also brought with it an urban revolution: technological innovations allowed industries to produce goods in large quantities and in short times, enjoying the benefits of the so-called economies of scale. The peasants quickly abandoned the tiring and patient work of the countryside to move to the cities, attracted by new opportunities for work and income: the cities therefore, in order to welcome more and more inhabitants, demolished the old medieval walls that delimited the perimeter to build new residential districts for the workers. The scenario at the end of the nineteenth century showed cities in rapid expansion and with a growing infrastructural and service endowment: at the same time, however, the neighborhoods remained poorly connected to the center and the hygienic conditions were often very bad.

In the contemporary era, cities take on their present form: from the 1920s onward, in particular, the production plants of large industries are transferred from the centers, where they were located in the previous century, to the more peripheral areas of the cities (and later even to other low labor costs—countries with the advent of globalization). Post-Fordist cities, freed from the physical presence of factories, reorganized their economy around tertiary services.

After the introduction of the concept of city and its historical evolution, it is now important to observe the role that these large poles play on the world stage, in economic, social, and environmental terms. As mentioned above, 2007 represents a historic year for humanity: for the first time, more people lived in cities than in rural territories (McDonald, 2008). The World Urbanization Prospects, drafted by the

United Nations in 2018, reports past data and future trends relating to both urban and rural population growth up to 2050. This study shows that the world urban population has grown by 3.47 billion in the period 1950–2018, going from just 0.75 billion people in the mid-twentieth century to 4.22 in 2018. Regarding the growth of the phenomenon of urbanization, in 1950 the population residing in cities (0.75 billion) was just about 30% of the total (2.54 billion), while in 2018 it almost doubled in percentage terms (55%) reaching 4.22 billion. This was made possible thanks to an urbanization rate (average annual rate of change of the urban population) equal to 0.92% on average every year in the period 1950–2018.

Furthermore, the UN report states that urban population reached 1 billion inhabitants in 1959, 2 billion in 1985 (26 years later), 3 billion in 2002 (17 years later), and 4 billion in 2015 (13 years later). According to current estimates, the phenomenon of urbanization should also continue in the future: cities should have 5 billion inhabitants in 2028 and 6 billion in 2041 (maintaining stable the number of years necessary to reach the next billion, i.e., 13). Although, as mentioned above, estimates foresee that the phenomenon of urbanization will continue in the future, it will do it at a slower rate than in the past: the growth of the urban population in the period 2018–2030 should be at 1.7% (a much lower rate than in the past) and even decrease to 1.3% in the following twenty years 2030–2050. On the contrary, the rural population reached 2 billion in 1960 and, 19 years later, 3 billion: estimates do not predict that it will reach 4 billion given the slow decline that began in 2021. Forecasts for 2050 estimate that the number of people residents in rural areas will reduce to 3.1 billion (lower than in 1990) (UN, 2018).

If abovementioned forecasts are correct, the growth of the world population up to 2050 will concern only people residing in urban areas and will not present a uniform distribution between cities: this phenomenon, which has in part already started, will involve the formation of mega-cities and peripheral centers. At this point, it is crucial to define the hierarchy of cities based on the number of inhabitants: at the top of the ranking of the largest settlements are the so-called mega-cities with more than ten million inhabitants (the 33 existing mega-cities collect 7% of the global population at 2018). Cities with a population between 5 and ten million are defined as "large cities" (4% of the world population), those with 1–five million inhabitants "medium-sized cities," while all the less populated agglomerations belong to the remaining categories (Slavova & Okwechime, 2016). Currently, the distribution of the population in cities is as follows: more than two thirds of the world's urban population lives in cities of less than 300,000 inhabitants, the remainder living in larger cities. Mega-cities are therefore distinguished by their size and the concentration of activities and services: in the last decade of the twentieth century, there were only 10 (153 million inhabitants in total) while in 2018 they more than tripled to 33 (529 million). Mega-cities are currently concentrated in only 20 countries, mainly in Asia (China hosts 6) but also in South America, Africa, and the United States. The UN forecasts to 2030 estimate the birth of an additional 10 mega-cities distributed among China, India, Africa, and South America, bringing the total to 43 (UN, 2018).

By 2030, therefore, cities will be bigger and with more inhabitants: this will have an unprecedented impact on their infrastructural endowment and resources. The ten most populous cities in the world (which will attract 35% of the population increase) are estimated to be Tokyo (37.2 million), Delhi (36.1 million), Shanghai (30.8 million), Mumbai (27.8 million), Beijing (27.7 million), Dhaka (27.4 million), Karachi (24.8 million), Cairo (24.5 million), Lagos (24.2 million), and Mexico City (23.9 million) (UN, World Urbanization Prospects: The 2018 Revision., 2018). Also the "smart cities," thanks to a high quality of life, attract flows of people who decide to seize the opportunities that this type of places offer. Winters (2011) in his study "Why are smart cities growing? who moves and who stays" analyzes migration flows toward "smart cities" in the United States: results of this study show that the main cause of migration is represented by the high quality of education. More and more young people are moving to "smart cities" to graduate, and then they often decide to stay and live in that city even at the end of their studies. The analysis also shows that most of these young people come from other cities within the same state, effectively generating an "intrastate brain drain" (Winters, 2011). In Italy, a study was conducted on the propensity of Italian citizens to migrate to a "smart city": 60% of them felt willing to move to live in such a city if it is located within their own region (PepeResearch, 2022). An urbanization phenomenon of this intensity concentrates most of the world's wealth (about 80% of GDP) in the metropolises, but also inevitably problems. These include worsening air pollution, chaotic urban expansion (such as the development of slums in certain areas of the world), inadequate and overloaded infrastructures and services, and an excessive consumption of energy resources (over half of the world's resources are absorbed by urban centers) (WorldBank, 2020).

An urbanization of such intensity generates areas in which tension, poverty, and pollution are concentrated, and for local administrations it will be increasingly difficult to manage urban fabric. Society on a global level, thanks to the increasingly widespread economic well-being, is observing a slow aging process: this phenomenon represents an important challenge especially for the cities of the world.

According to the Organization for Economic Cooperation and Development (OECD), 43,2% of the global population over 65 resides in cities (OECD, 2015). Local administrations must analyze this trend and plan appropriate measures in time. If on the one hand the lengthening of the average life span of people represents a positive factor due to the economic and technological progress of society, at the same time cities must be able to guarantee to the elderly maximum social inclusion and accessibility to urban spaces (van Hoof & Kazak, 2018). The World Health Organization (WHO) in 2007 suggested to cities to adopt an "age-friendly cities" model by adopting some measures aimed at protecting the oldest component of their population (Buffel et al., 2012). According to WHO, cities must maintain "active aging by optimizing opportunities for health, participation and security in order to enhance quality of life as people age" (WHO, 2007).

Based on the above, the cities of the world must redefine their welfare through the involvement of all the stakeholders: in a historical context devoted to flexibility and adaptability, a rigid welfare state would be anachronistic in the face of the profound changes taking place. The ideal welfare of the cities of the future must be able to recognize and satisfy the needs of the population and stimulate the competitiveness of the current economic model without compromising the protection of citizens' rights. The problems that cities face in this historical period, if badly managed by the administrations, can generate social inequalities that over time lead to civil tension: the construction of a form of urban resilience becomes fundamental in view of the social and economic development of the city. The Open Working Group on Sustainable Development Goals, held at the UN glasshouse, recently defined important intervention measures to build a more sustainable and balanced world: in September 2015, the governments of 193 countries signed a joint program which drafted 17 Sustainable Development Goals (SDGs) to be achieved by 2030 (Allen et al., 2018). Among these 17 objectives, 2 relating to cities emerge: objective 11, "Sustainable cities and communities," specifically concerns urban development and aims to make cities and human settlements more inclusive, safe, and resilient. Goal 15, on the other hand, "Life on land," aims to protect and promote sustainable use of the terrestrial ecosystem in urban and non-urban areas. As for the Sustainable Development goal number 11, dedicated specifically to the theme of cities, it aims to ensure equitable access to housing and to redevelop slums, provide safe; inclusive, and sustainable access to urban public transport; protect cultural heritage of cities; decrease the number of deaths directly related to economic problems; reduce the per capita environmental impact of cities; improve urban waste management; etc. (UN, 2022). In relation to goal 15, dedicated to life on land, the so-called "nature-based" solutions (NBS) are spreading around the world facing the problem of the growing imbalance in terrestrial ecosystems: NBS are interventions that use what nature offers to improve the quality of life of citizens both in urban and non-urban areas.

In a 2015 report, the European Commission defined NBS as follows:

> Solutions that are inspired and supported by nature, which are cost-effective, simultaneously provide environmental, social and economic benefits and help build resilience. Such solutions bring more, and more diverse, nature and natural features and processes into cities, landscapes and seascapes, through locally adapted, resource-efficient and systemic interventions. (EU, 2021).

In urban contexts, concrete examples of NBS are those interventions that make city neighborhoods greener, bringing social and psychological benefits to citizens: e.g., introduction of green roofs and walls, urban gardens, urban oases of biodiversity, etc. The advantages of this type of measures concern the mitigation of the urban climate (reduction of heat islands), limitation of the emission of harmful substances, protection and increase of biodiversity, reduction of the risk of flooding due to the water absorption of plants, less acoustic pollution (the plants act as sound-absorbing

Table 6.1 EU NBS projects. Source: authors' elaboration based on (cordis.europa.eu, n.d.)

Project	Brief description	Urban
CONNECTING Nature	The project aims to form a community of cities for the exchange and mutual learning of know-how related to NBS: leading cities (with experience in NBS already applied) and gregarious cities (who want to learn) belong to this cluster. A total of 11 cities from Belgium, Bosnia and Herzegovina, Bulgaria, Cyprus, Greece, Italy, Poland, Scotland and Spain are involved	√
GROW GREEN	The project aims to help cities achieve lasting change by integrating nature-based solutions into their planning, development, and management (e.g., through the creation of green spaces and waterways). The cities involved are Manchester, Valencia, and Wroclaw	√
URBAN GreenUP	The project demonstrates ecological solutions aimed at reforming urban areas and increasing their sustainability and resilience to climate change. The project involves 25 partners from 9 different EU countries	√
NATURVATION	The project, in collaboration with the ICLEI institute, seeks to provide new methods for successfully implementing NBS to cities	√
UNALAB	The project aims to provide a wide range of tools and methodologies to improve the water resilience of cities: e.g., wetlands and water retention ponds for thunderstorms. The cities involved at the moment are Genoa, Eindhoven, and Tampere	√
Nature4Cities	The project promotes collaborative models for the adoption of NBS, it has also developed an online platform for sharing and supporting decisions	√
ThinkNature	The project has created a communication platform between stakeholders to support and promote NBS initiatives at local, regional and European level. It currently involves 126 partners and international organizations	√
MERCES	The project aims to restore marine ecosystems in European seas and international waters, making them resilient to climate change	×
NAIAD	The project aims to provide NBS to counter the risks of floods and droughts in cities and river basins	√

barriers), and energy saving (the plants reduce the need for home heating in winter and air conditioning in summer) (EEA, 2021). The European Commission has been very attentive to the NBS issue and has decided to finance nine different related projects, briefly described in Table 6.1.

In an increasingly urbanized and globalized world, the winning choice will go through the ability to manage consciously the limited resources available and to create a cluster, made up of the population, companies, and institutions, which favors the participation and involvement of all stakeholders in public life in an equitable manner.

6.2 The "Digital City"

The profile of the city of the future described above can take different forms depending on the technological tools used and the degree of citizen involvement: the first step in the evolutionary process of a city is represented by its degree of Internet connection. The so-called digital city provides a multitude of sensors scattered around the urban environment capable of collecting data and information released by citizens: the data are then managed and processed by technological platforms supplied to the city administration. Thanks to this instrumentation that allows to fully understand the needs, movements, and attitudes of citizens, cities are able to offer more efficient services (Laguerre, 2005).

Ishida (2002) argued that the main types of technologies of a "digital city" are the following: technologies for integrating information (crucial for collecting data and then managing and processing them through technological platforms), technologies for involving the population (allowing citizens to take an active part in the creation of the new "digital city"), and technologies for social agents and technologies for information security (cybersecurity essential to prevent cyber attacks). In a "digital city," shared mobility is the key element: vehicle-charging stations are scattered throughout the city in order to integrate public and private mobility. The connection to the building network is also an important step: they are connected to the urban smart grid to allow the administration to constantly monitor consumption. This city paradigm has thus been made possible by the increasing global deployment of Information and Communication Technology (ICT) services, which have also found application in the urban environment since the 1990s: technology has proven to be an essential condition for sustainable urban development.

6.3 From "Digital Cities" to "Smart Cities"

The evolutionary step following the "digital cities" is represented by cities that absorb technological innovations in the urban planning phase and in the management of urban infrastructures: the "Smart Cities" (Dameri, 2013). A univocal definition of "smart city" does not exist and over time academic literature provided various interpretations: Table 6.2 shows the definitions offered by some of the most important institutions on a global level (Lai et al., 2020).

Based on the definitions in the table above, it is possible to find similarities and differences: this synthesis work allows to fully understand what is meant by "smart city." The aforementioned definitions agree in defining "smart cities" as spaces in which it is possible to use ICT services to improve the quality of life of people through a process of data collection and elaboration, urban systems are integrated, and the main objective concerns the development of economic and social sustainability. Although the principles of "smart cities" are common to all stakeholders, the sectors (e.g., transport, energy, health, etc.) with the greatest urgency for intervention

Table 6.2 Main "Smart City" definitions. Source: authors' elaboration based on (Lai et al., 2020)

Organization	Definition
Association of Southeast Asian Nations	"A smart city in ASEAN harnesses technological and digital solutions as well as innovative non-technological means to address urban challenges, continuously improving people's lives and creating new opportunities. A smart city is also equivalent to a "smart sustainable city", promoting economic and social development alongside environmental protection through effective mechanisms to meet the current and future challenges of its people, while leaving no one behind. As a city's nature remains an important foundation of its economic development and competitive advantage, smart city development should also be designed in accordance with its natural characteristics and potentials".
British Standard Institution	"A smart city is an effective integration of physical, digital and human systems in the built environment to deliver a sustainable, prosperous and inclusive future for its citizens".
Department for Business, Innovation and Skills, UK	"A Smart City should enable every citizen to engage with all the services on offer, public as well as private, in a way best suited to his or her needs. It brings together hard infrastructure, social capital including local skills and community institutions, and (digital) technologies to fuel sustainable economic development and provide an attractive environment for all".
European Commission	"A smart city is a place where traditional networks and services are made more efficient with the use of digital and telecommunication technologies for the benefit of its inhabitants and business. A smart city goes beyond the use of ICT for better resource use and less emissions. It means smarter urban transport networks, upgraded water supply and waste disposal facilities and more efficient ways to light and heat buildings. It also means a more interactive and responsive city administration, safer public spaces and meeting the needs of an ageing population".
Innovation and Technology Bureau, Hong Kong	"Embrace innovation and technology to build a world-famed Smart Hong Kong characterized by a strong economy and high quality of living".
Institute of Electrical and Electronics Engineers Smart Cities Community	"A smart city gathers government, technology, and society to achieve a minimum of the following factors: smart mobility, a smart economy, a smart environment, smart cities, smart governance, smart people, and smart living."
International Electrotechnical Commission	"A smart city is one where the individual city systems are managed in a more integrated and coherent way, through the use of new technologies and specifically through the increasing availability of data and the way that this can provide solid evidence for good decision making".

(continued)

Table 6.2 (continued)

Organization	Definition
Japan Smart Community Alliance	"A smart community is a community where various next-generation technologies and advanced social systems are effectively integrated and utilized, including the efficient use of energy, utilization of heat and unused energy sources, improvement of local transportation systems and transformation of the everyday lives of citizens"
Ministry of Housing and Urban Affairs, India	"The conceptualization of Smart City, therefore, varies from city to city and country to country, depending on the level of development, willingness to change and reform, resources and aspirations of the city residents. A smart city would have a different connotation in India than, say, Europe. Even in India, there is no one way of defining a smart city".

depend on specific regional interests (Lai et al., 2020). Within a "smart city," in comparison with a common city, citizens can benefit from better protection for their investments, more efficient traffic management (through, e.g., smart traffic lights), powerful connectivity capable to attract and develop new businesses, a conspicuous reduction in energy consumption and associated costs, powerful and free Wi-Fi connection, sustainable, and shared mobility (Bibri, 2018). Furthermore, according to the literature, a "smart city" must undertake ecological measures such as the following: reuse of landfills biogas to produce new and clean energy, construction of buildings with low environmental impact, decrease in dependence on fossil sources through investments in renewable energy, and reduction of digital divide among citizens. Tang et al. (2019) conducted a study on 60 global cities municipal "smart city plans" in order to identify the main programs and policies implemented. From this work, the authors identified four different models of "smart cities" currently existing in the world:

- *Essential services model*: this type of model is adopted in particular by cities (such as Tokyo and Copenhagen) which, in the management of emergencies, intend to heavily invest in the use of telecommunications and digital healthcare services.
- *Smart transportation model*: a model of this type is suitable for heavily populated cities that have serious problems related to traffic management. For this reason cities like Singapore and Dubai have adopted this smart city model through investments in sustainable and smart public transport, smart traffic lights, and shared and autonomous vehicles.
- *Broad-spectrum model*: cities such as Beijing and Barcelona have adopted this smart city model in order to use intelligent solutions for the management of urban services (e.g., water and waste management and pollution reduction).

- *Business ecosystem model*: this smart city model envisages huge investments in digital skills training programs in order to create a qualified workforce for high-tech companies (e.g., Amsterdam and Edinburgh).

In order to fully materialize and develop, smart cities need the participation of all the stakeholders involved: institutions, companies, and citizens. The population in particular is required to have a proactive attitude in order to be an integral and active part of the city change project: only in this way the measures launched by the institutions can take root in the social fabric of the city. In this regard, it is crucial that citizens are well aware of the change taking place and willing to be part of it: the change must start from the mentality of the people and not from the availability of technologies. In 2020 the Capgemini Research Institute thus decided to submit a questionnaire to the urban population (10,000 citizens and more than 300 municipal officials involved belonging to 58 cities around the world) to assess whether cities meet the expectations of citizens as they are or whether people are eager of cities more digital advanced. The results of the survey converged in July 2020 in the report "Street Smart: putting the citizen at the center of smart city initiatives" which stated that about 40% of residents are disappointed in their city due to a lack of digital development. The report analyzed the citizens' so-called willingness to pay: a third of respondents (36%) were willing to pay more to get a better urban experience than the current one. This trend increased if the categories of respondents were younger and wealthier (Capgemini, 2020). 58% of citizens considered "smart cities" to be sustainable and ensuring a better quality of the services (57%). The cities of the world did not result from the report at a good conversion stage toward the concept of "smart city": only 10% of officials believed their city was in an advanced stage of conversion while just 22% of cities started to take measures to become a "smart city." The report showed how urban citizens increasingly consider environmental sustainability a priority: they were concerned about how their city manages the challenge of pollution (42%) and sustainability in general (36%) (Harper, 2020).

The aforementioned data reveal that citizens are in favor of the change toward a new type of city that integrates ICT services in all its aspects: results of this kind show that citizens are ready to do their part in the transition to intelligent and digital cities. To become smart, a city must place knowledge at the center of its strategic vision: such an approach to urban planning falls under the term of "Knowledge-Based Urban Development" (KBUD) (Yigitcanlar, 2010). The KBUD concept allows cities to make the transition to more sustainable and inclusive places: cities' goal of attracting and retaining human and intellectual capital has a positive effect on the development of their territory at an economic, social, and cultural level (Pancholi et al., 2015). Penco et al. (2020) analyzed the relationship between knowledge-based urban environment and entrepreneurship through the development of a multidimensional index called KBCDE (Knowledge Based City Developing Entrepreneurship): the study, conducted on 60 European cities, showed the relationships between the KBCDE index and some selected sub-indixes (social and talent-cultural perspective, economy and context economy perspective, environmental and infrastructural perspective, urban innovation system perspective). The results allowed the cities to be clustered into three different groups on the basis of the greater degree of "knowledge city" and to provide some indications to policymakers: urban policies aimed at increasing the human and intellectual capital of the city also stimulate the entrepreneurial sector, referring to attraction of new and digital companies.

6.4 From "Smart Cities" to "Sustainable Cities"

If the key element of "Smart Cities" was represented by the extensive use of technology in city management, the main objective of "Sustainable Cities" concerns the monitoring and reduction of the environmental impact of cities (Matos et al., 2017). The measures to make a city sustainable are different and all aimed at increasing the quality of life of citizens without compromising the health of the environment. The mobility of citizens represents an important step in the change toward a smog-free city: the main measures taken in this regard concern the replacement of public transport vehicles with zero-impact vehicles (electric or hydrogen in the future), construction of pedestrian areas and cycle lanes, renewable solar and wind energy plants, and reduction of food and water waste. For example, Akande et al. (2019) have drawn up a ranking of the 28 capitals of the European Union based on their degree of sustainability and intelligence: the list has been constructed taking into account 32 different indicators grouped into 4 components. Ciacci et al. (2021) instead concentrated on the Italian provinces: their analysis aimed at drawing up a ranking of cities on the basis of how "smart sustainable" they were. To conduct this study, a partially non-compensatory quantitative method (called Pena's distance (DP2)) was used, which made it possible to compare the performance of cities taking into account a mix of functional elements characteristic of each area. These distinctive elements, whose data were found on the ISTAT[1] database, referred to six different macro-categories: environment, mobility, technology, economy, education, and culture. This work has shown how Siena, Milan, and Padua rank at the top in Italy as the closest cities to being "smart sustainable"; on the contrary, the cities of Carbonia, Barletta, Caltanissetta, Trapani, and Andria obtained the worst results.

Bibri and Krogstie (2017) conducted an extensive literature review on the concepts of "smart" and "sustainable" cities. The authors, after carefully analyzing the academic literature, reported the most widespread definitions, the state of the art, and the possible future developments of the two paradigms: the results of the study showed that the main problems associated with the adoption of such concepts remain largely under-explored by academics. Furthermore, the authors, aware that each city deals with the issue independently, have developed a framework for sustainable and smart urban development that takes into account the main weaknesses and strengths of the models examined. Cities need a medium-long-term strategic vision with a

[1] Italian National Institute of Statistics.

view to greater environmental and economic sustainability. In this sense, Gothenburg has recently developed the initiative "The Project Gothenburg 2050" which aims to transform the Scandinavian city into a definitely "smart and sustainable" place. The methodology underlying this measure is very simple, usable for any sector and essentially composed of four steps: description of the state of the art, definition of the main sustainability objectives to be achieved, understanding of what type of city to become, and finding a way to become that kind as soon as possible (Bibri, 2018).

Ivaldi et al. (2020) and Penco et al. (2019) finally studied the relationship between SSC and "Urban Knowledge-Based Economy": the main purpose of the analysis was to understand whether a SSC environment can create the conditions suitable for the development of the urban knowledge-based economy and what are the most decisive domains in doing this. The study, conducted on 116 Italian cities, used multidimensional indexes to describe the different SSC domains and how much they influence the knowledge-based economy. The results of the analysis showed that a sustainable smart city creates an environment suitable for the urban knowledge-based economy: in particular, thanks to the development of sustainable social services and sustainable local transport.

6.5 From "Sustainable Cities" to "Resilient Cities"

The most important challenge of the twenty-first century concerns climate change: cities must not only be digital, smart, and sustainable, but also resilient to extreme weather events. A negative example of mega-cities unable to cope with the challenges of global warming is represented by Jakarta, the capital of Indonesia: as a result of chronic congestion, pollution at record levels, and increasingly frequent flooding, the land on which the city stands is sinking 25 cm every year. The country's government has thus recently taken the radical decision to move the city to a new, safer area: the project, which is expected to cost $ 32 billion, aims to build the new capital Nusantara in a remote area of Borneo with devastating impacts on the local ecosystem (Rahmat et al., 2021). In order to avoid what happened in Indonesia, the cities of the world must plan countermeasures very quickly. The concept of "resilient city" has gained increasing attention since the devastating effects of Hurricane Sandy: in 2012, it struck with extraordinary force on the northeastern coast of the United States (New York in particular), causing the death of 43 people and economic damage for about $ 19 billion (Papa et al., 2015). The "resilient city" must develop a high degree of adaptability to the social, economic, and environmental changes taking place in order to guarantee lasting and stable prosperity over time (Bruzzone et al., 2021). As for "smart cities," even the definition of "resilient city" is not unique at a global level: Papa et al. (2015) collected the main definitions found in academic literature, in Table 6.3.

Compared to the last century, the largest cities in the world today represent real complex urban systems with a series of intertwined relationships: a scenario of this

Table 6.3 Main "resilient city" definitions. Source: authors' elaboration based on (Papa et al., 2015)

Reference	Definition
Newman et al. (2009)	*Resilient cities have built-in systems that can adapt to change, such a diversity of transport and land-use systems and multiple sources of renewable power that will allow a city to survive shortages in fuel supplies.*
Kourtit et al. (2012)	*A resilient city is also a creative city, able to reinvent a new equilibrium against destabilizing external pressure. It multiplies the potential of people to build new opportunities/alternatives.*
Pickett et al. (2004)	*Resilient City is a city that supports the development of greater resilient in its institutions, infrastructure and social and economic life. Resilient cities reduce vulnerability to extreme events and respond creatively to economic, social and environmental change in order to increase their long- term sustainability.*
UNISDR (2012)	*A resilient city is characterized by its capacity to withstand or absorb the impact of a hazard through resistance or adaptation, which enable it to maintain certain basic functions and structures during a crisis, and bounce back or recover from an event.*

type exposes cities to unprecedented risks. Local administrations have the priority of drawing up urban risk management and mitigation plans: a defense program based on contrasting one risk at a time is no longer considered possible, but an integrated and structured approach to risks must be developed, making cities invulnerable and resilient. There are many projects globally (e.g., 100 Resilient Cities,[2] C40,[3] IPCC[4]) aimed at increasing the resilience of cities through a reorganization of sectors such as mobility, water and waste management, infrastructure design, etc. An international virtuous example is represented by the city of Yokohama in Japan: the high population density of the metropolis exposes it to numerous risks. Therefore, the administration has decided to implement various initiatives aimed at making it more sustainable and resilient to external attacks. In particular, the mayor aims to promote the use of renewables, manage sewage sludge and waste in a more efficient and ecological way, and develop sustainable mobility (TDLC, 2017). Rotterdam, on the other hand, is a city located below the level of the Atlantic Ocean, and this exposes it to a very high risk of flooding (Lu & Stead, 2013): in 2007 an ambition urban reorganization program was launched aimed at protecting the city from extreme weather events. Specifically, the so-called water squares were built: open spaces furnished with sports machinery and benches capable of absorbing large quantities of water, thus avoiding flooding the rest of the city. Melkunaite and Guay (2016) reported in their study on the relationship between urban planning and civil protection in designing of resilient urban planning programs that "resilient cities" are able to respond to external attacks thanks to their own conformation: in fact, they present

[2]"100 Resilient Cities" project has been pioneered by the Rockefeller Foundation in order to help cities around the world to achieve a better resilience to external shocks.

[3]C40 is a group of 97 cities around the world with the goal of fighting climate change.

[4]Intergovernmental Panel on Climate Change (IPCC) is the most important international organism aimed at evaluate climate change.

elements of redundancy, capacity for reorganization, flexibility, and the ability to learn the lessons of history. Bruzzone et al. (2021) focused on reporting activities related to the resilience of cities: this practice is fundamental for administrations as it allows them to monitor the results obtained over time and the effectiveness of their actions aimed at strengthening urban resilience. Their work was based on the identification of the main elements necessary for the preparation of a city resilience reporting: to this end, they investigated the urban resilience policies of the cities of Barcelona, London, Athens, and Lisbon.

6.6 From "Resilient Cities" to the "15 Minute City"

One of the main obstacles to the complete realization of sustainable cities is represented by the practice of "urban sprawl." This phenomenon relates to an urban planning, typical of the past decades, which provides for the unlimited expansion of the city beyond its suburbs. A situation of this kind causes a great environmental impact due to the necessary transport infrastructures to be built, the sewer networks, the disorientation and expropriation of the fauna that lived in those areas, and the loss of flora. The cities of the future, in order to reduce polluting emissions, will have to be built in such a way that people have all the essential services readily available: one of the most innovative solutions to overcome this problem is the so-called 15 minute city. Initially proposed by Carlos Moreno, Scientific director of Chair "Entrepreneurship—Territory—Innovation," Panthéon Sorbonne University—IAE Paris (France), on the occasion of the electoral campaign of the mayor Anne Hidalgo in Paris, it consists of an ideal urban space in which citizens can satisfy their essential needs such as going to work, to the hospital, to school, to the gym, or shopping with trips on foot or by bicycle, taking up to 15 min (Pinto & Akhavan, 2022; Moreno, 2020). The main goal of an urban project of this kind is to allow all citizens to get access to everything they need without using the car, thus promoting a healthy, active, and environmental-friendly lifestyle (Moreno & Ochoa, 2011). Carlos Moreno intends to rethink what planners and architects have always tried to achieve in the city: that is, to allow citizens to reach two points A and B distant in space in the shortest possible time. The intent of his project is diametrically opposed and aims to eliminate the need to make these polluting and energy-intensive movements, preferring proximity to essential services. Through the concept of "15-minutes city," citizens are able to regain possession of their time without having to spend hours and hours stuck in traffic due to long journeys (Allam et al., 2021). The roads, used nowadays only by cars, are redesigned according to people needs: Space occupied by cars is replaced by parks, fountains, and vegetation that mitigate the heat island effect and lower temperatures.

In Paris, as mentioned, this ecological planning of the city raised the enthusiasm and allowed the mayor Anne Hidalgo to be reconfirmed in the municipal elections: after that, the city began to see what was planned into reality. Taking advantage of the COVID-19 period in which transport had been greatly reduced due to the restrictions imposed by the pandemic, cycle paths and green spaces were gradually introduced in the streets of central Paris to make the city suitable for pedestrians and cyclists. The second step in the program will be to transform the Parisian arrondissements into real "proximity neighborhoods." The case of Paris is under everyone's attention, and many cities around the world are starting to plan the same measures for their urban context. With the main goal of decongesting the road arteries during rush hour and aiming to reduce the phenomenon of work commuting, Milan is the first city in Italy that intends to make the Parisian project its own, while in Spain it is the city of Barcelona that is carrying on Moreno's ideas. From the 2013 urban mobility plan, the intention of the local administration is to create the so-called superblocks: predominantly pedestrian neighborhoods that constitute different small communities connected to other superblocks by external traffic routes (Rueda, 2019). Mueller et al. (2020) showed how superblocks in Barcelona work: the streets inside the superblock are used for soft mobility or for the exclusive use of residents with a maximum speed of 20 km/h, while the road around the perimeter allows for higher speeds and is free to use.

6.7 From the "15 Minute City" to the "Circular City"

A step beyond the "15 minute city" is that of the so-called circular city: according to the definition of the Ellen MacArthur Foundation, they are agglomerations that absorb all the principles of the circular economy (EMF, 2019). The economic model prevailing today is still linear and has been in operation since the industrial revolution: this concept involves the extraction of raw materials, their transformation into a finished product, their sale on the market, and their disposal as a waste. This approach, academics realized around the middle of the last century, causes a devastating impact on the environment because of the uncontrolled disposal of waste and the intensive exploitation of limited resources that cause their exhaustion. One of the first economists to denounce the unsustainability of this type of economic model was Kenneth Boulding in 1966 with "The economics of the coming spaceship Earth": the British economist used the figures of cowboy and astronaut to represent two opposing economic models. Boulding compared the linear (or open) economy to the conquering attitudes of the cowboy, who, eager to conquer west America, exploits resources in an uncontrolled way considering them to be infinite. On the contrary, the nascent concept of the circular (or closed) economy belongs to the astronaut who, forced to live in the closed space of a spaceship, uses resources in a thoughtful way, well aware of their limitation and preciousness. Boulding's contribution was decisive in offering an alternative solution to the historical linear economic model: only recently, however, as a possible response to the growing negative externalities of climate change, the concept of circular economy has been gaining increasing attention and importance for decision-makers all over the world. In 2009 the Ellen MacArthur Foundation was founded and immediately placed the transformation of the economic model from linear to circular at the core of its

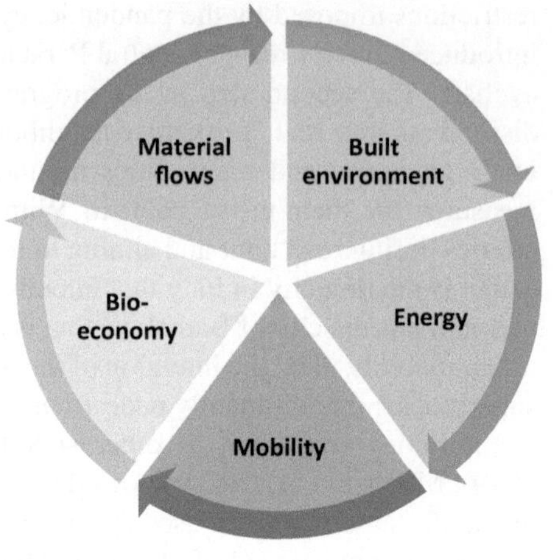

Fig. 6.1 The five pillars of a "circular city." Source: authors' elaboration based on (Meini et al., 2019)

mission: the actions taken concern the supply of tools for education, the development of studies and research, and the involvement of private companies and institutions in the project. In 2015, the European Union also drafted 54 actions, aimed at favoring the transition toward a circular economy, contained within the first package of measures specifically dedicated to it. Finally, in 2021, the European Parliament also approved the "Circular Economy Action Plan" with which the EU undertakes to follow this change.

Cities represent the ideal places to apply the principles of the circular economy: the high concentration of resources, capital, data, and people in a limited space generates fertile ground for the implementation of circular business models, reuse systems, and products-as-a-service models (EMF, 2019). A "circular city," through the concept of "3Rs" (reduce-reuse-recycle), aims to increase its resilience and the quality of life of citizens: these cities aim to eliminate waste through the reuse of materials in an optics of complete circularity in the life of the products (Fujita, 2009). According to Meini et al. (2019), there are five key urban subsystems to invest to obtain a "circular city": infrastructures, renewable energy, transport, bio-economy, and material flows such as waste, water, and food (Fig. 6.1).

Three are the pillars of a circular city where administrations must invest: buildings, mobility, and products. As far as buildings are concerned, one of the phenomena to be stopped concerns the abovementioned "Urban Sprawl" practice: continuing to extend cities beyond the suburbs involves, as mentioned, significant negative externalities on the environment and huge costs for local administration. Due to the long distances between the suburbs and the center, citizens have to travel several kilometers every day to reach places for jobs, study, or leisure. This involves long and frequent journeys that ask for huge maintenance costs on the transport arteries and harmful emissions into the atmosphere (Johnson, 2001). Mobility represents another fundamental step in the transformation of cities from linear to circular: the use of the sharing mobility concept allows people to purchase fewer vehicles, reduce the occupation of urban land by cars and contain harmful emissions. Finally, the consumption model of the products must be entirely reorganized according to the 3R logic: the waste materials must be reconditioned to be used for other uses.

In 2017, the book "Doughnut Economics: Seven Ways to Think Like a 21st Century Economist" was published, in which Kate Raworth proposes an alternative and sustainable economic model to the current one. The authors analyzed economic history and soon realized that the classic dictates of economics, the pursuit of infinite growth, have devastating side effects: exhaustion of the resources available on the planet, generation of economic and social inequalities, and enormous consequences in environmental terms. Raworth thus theorized a doughnut-shaped economic model, in which at the center (in the doughnut hole) are placed the basic needs of people (housing, networks, energy, water, food, health, education, income and work, justice, political voice, social equity, gender equality) while outside there are ecological limits not to be exceeded (climate change, ocean acidification, chemical pollution, nitrogen and phosphorus increase, freshwater withdrawals, land conversion, biodiversity loss, air pollution, ozone layer depletion). On the dough of the doughnut, between the ecological ceiling and the social base, people can live in an equitable and sustainable way: this space guarantees citizens respect for all fundamental rights and the necessary protection for the environment (Raworth, 2017a, 2017b). The first city in the world to apply this economic model to its urban planning was Amsterdam: the Dutch capital has set itself the goal of halving the use of raw materials by 2030 and becoming fully circular by 2050. Furthermore, the city intends to produce items with materials suitable for reuse, develop sharing economy platforms to give a second life to consumer products, fight food waste, and build buildings with sustainable materials (Maldini, 2021).

6.8 Discussion and Conclusions

The current historical context presents many challenges for the cities of the world: increasing urbanization, climate change, congested transport infrastructures, high number of road accidents, and degraded suburbs. To these elements was added the COVID-19 pandemic in 2020 which completely upset the habits of citizens and forced the administrations to take extraordinary measures to contain the virus through a social distancing that was not always possible: in many mega-cities, in particular in Asia, it can be really complex to guarantee the safety of citizens due to the overcrowding of urban spaces. The pandemic has also brought structural changes to the way of working: due to the restrictions imposed by governments on the free movement of people, particularly in the most acute phases of the contagion, the only possibility to continue working turned out to be remote work. More and more companies then, discovering the economic benefits that this work model entails, have sometimes decided to make this practice permanent even after lockdowns (or in any case to guarantee more flexible working hours compared to pre-pandemic times).

This historical phase has also seen the definitive boom of e-commerce phenomenon: in this period people, to limit movements and infections as much as possible, enormously increased their online purchases. The urban logistics sector was put under stress and had to adapt its business structure to the new large amount of work required: the consolidation of this trend, even after the acute phases of the pandemic, involves negative externalities such as congestion of road arteries and large emissions of CO_2.

Cities are therefore faced with numerous problems that make their management very complex. The purpose of this paper was to show the need for a profound change in the structure of cities, which is fundamental in order to respond adequately to current challenges. The theme of the city of the future in all its forms has been extensively treated in academic literature. Each new city paradigm ("digital," "smart," "sustainable," "resilient," "15 minute," "circular") is treated individually and independently from the others: very little is, conversely, to be found on the relationships between the different evolutionary phases. For this reason, this work, starting from the assumptions made by Bruzzone et al. (2021) in "Resilience Reporting for Sustainable Development in Cities," aims to fill this gap by offering an overview of the evolutionary phases of the cities and by analyzing the relationship that binds them.

The evolutionary steps of the city of the future, as described below, all depend on a single key element: technology. The city can take on different forms and evolve over time based on the degree of use of technological innovations and their penetration into the social, economic, and environmental fabric. The first concept of the city of the future was the so-called digital city in which the goal of administrations was a wide connection of the city to the Internet: the myriad of information that citizens release every day are collected by special technological platforms that, processing these data, are able to suggest to policymakers which services are most requested and needed by the population. Digitization, the preliminary step to the concept of "smart city," allows cities the centralized and coordinated management of four fundamental urban aspects such as mobility, sustainability, culture, and services. If the "digital city" does not provide for the direct involvement of citizens, the next step is taken by the "smart city": after being fully connected to the web, cities must use technological innovation directly in urban planning and infrastructure management and offer high quality services to citizens based on their needs (Dameri, 2013). Higher quality services attract flows of people who seek better job and study opportunities in these cities: as mentioned above, the main migratory flows toward "smart cities" concern the category of young students looking for a study offer of high quality. Often the young people, thanks to the better life conditions in the city, remain to live in that place even at the end of the cycle of studies: urban administrations must prepare adequate urban plans to cope with a constant increase in urban population. With "smart cities," the life quality of citizens begins to rise: the environmental aspect is only partly included in this scenario. For this reason "sustainable cities," the next level in the evolutionary scale of cities, exploit the most recent technological innovations to make the urban fabric eco-sustainable: reduction of atmospheric and acoustic polluting emissions, production of clean energy from renewable sources, construction of buildings with green materials, etc. (Bătăgan, 2011). The "resilient city," next step to the "sustainable city," combines all the aspects of the previous paradigms in a single model: resilient cities use

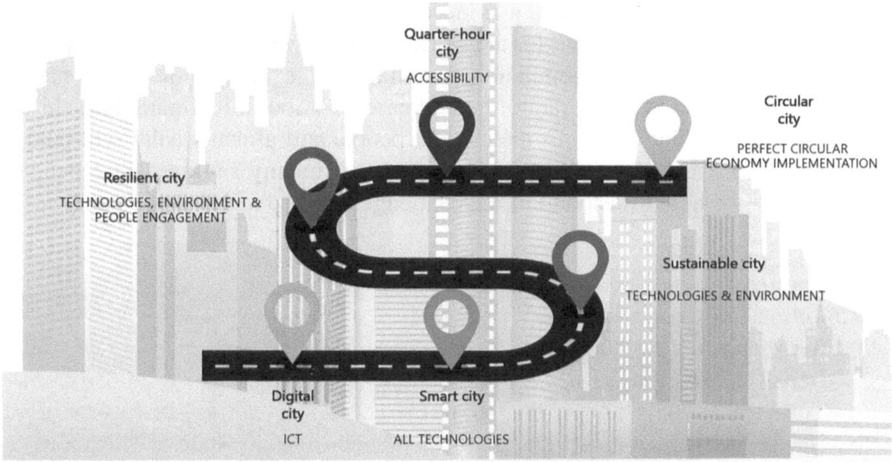

Fig. 6.2 Cities evolution over time

technology to cope with unforeseen events and sudden shocks, avoiding damages to their social, physical, and economic urban fabric. To do this, administrations need to build strong and cohesive communities: the key element of this model is represented by the strong involvement of citizens in the drafting and communicating phases of urban planning. A resilient city is a city that has been able to involve, train, and stimulate its own population. Bruzzone et al. (2021) stop at this point while the present study intends to add two further steps in the evolution scale of cities: the "15 minute city" and the "circular city." The concept of "15 minutes city" conceived by Carlos Moreno can be placed immediately after the "resilient city" model: de facto the two paradigms share the same innovative elements with the use of urban space as a substantial difference. The "resilient city" applies its model to the entire urban territory without distinction while the "15 minute city" prioritizes the reorganization of space in order to limit the movement of the population to a radius of 15 minutes from their home: this model involves a considerable improvement in the life quality of people due to the creation of a city on a human scale, a lower environmental impact, and a redevelopment of the most degraded areas. The final step, at the moment, is represented by the concept of "circular city." A key element of the "15 minutes city" paradigm concerns the general reorganization of consumption: people, having everything they need available in their neighborhood, no longer have to make long trips to make purchases. Vice versa, the "circular city" intends above all to redefine the production phase: the materials used for the production of buildings, infrastructures, vehicles, and objects are chosen on the basis of their degree of reusability. Furthermore, a city of this kind, with a view to conserving resources and reducing waste, intends to halt the uncontrolled urban expansion beyond its periphery through the redevelopment of existing neighborhoods: this paradigm can be applied to a city as a whole or even better to the "15 minutes city" model. The circular city is an equitable city that offers its citizens a high quality of life without compromising the sustainability of the environment. Figure 6.2 illustrates the different evolutionary phases of a city as described above.

In conclusion of this chapter, it is necessary to underline the importance that urban changes of this type entail for the world in which we live. The problems of our time (climate change, overpopulation, malnutrition, overexploitation of natural resources, air pollution, etc.) cannot be addressed without a real paradigm shift at the urban level. In the future, as mentioned, people and global wealth will tend to concentrate more and more in cities: a large part of humanity's future will depend on how they will be able to respond to the challenges of the twenty-first century.

Singapore
Country: Singapore
Inhabitants: 5,637,000
Density: 7804/km^2 (20,212.3/sq. mi)
Total area: 733.1 km^2 (238.1 sq. mi)

One of the world's most relevant financial hubs, the city-state of Singapore, located south of Malaysia, aspires to evolve into a truly Smart Nation in the future. The two elements that have supported the dizzying economic growth of the metropolis from the mid-twentieth century to today are essentially attributable to a large workforce and the attraction of investment capital: with the general aging of the population and the slowing of the immigration process due to the limitation of space, the goal to be pursued in order to continue economic growth is total factor productivity, which can be achieved through the full application of technology in all aspects of society.

Singapore is one of the world's most densely populated cities, and because it is an island, it is impossible to expand outside of its own territory. As a result, the government has been forced to come up with circular solutions to keep the city habitable, sustainable, and less import-dependent. The concept of circularity was first put into practice in the building industry with the creation of the so-called green buildings: some of which have already been built with recycled materials such as wood while the roofs of others have been covered with vegetation capable of absorbing carbon dioxide and lower tropical temperatures.

Some policies also involve stringent protocols for the reuse of materials gained through building demolition: the effectiveness of this measure has persuaded the local government to extend this recovery procedure to organic, packaging, and electronic waste as well. For food, on-site waste treatment systems that can turn food scraps into compost for the grounds are increasingly common in restaurants and residential complexes. In terms of product packaging recycling, Singapore has already been active for several years in training activities at producer companies and awareness raising at schools and workplaces. The city-state has resolved to establish specialized collection locations for electronic garbage in every neighborhood and recently ordered manufacturing firms to be responsible for the end-of-life handling of products.

(continued)

The Singaporean government takes the issue of garbage recycling seriously since, based on projections, the city's sole landfill should be full by 2035. To boost the rate of products recycling, particularly plastic ones, a Zero Waste Masterplan has been advocated. The degree of digitalization of Singapore's government, business, and society as a whole will determine how quickly it becomes a completely smart and circular metropolis.

Although Singapore's qualities as a city-state make it look like a case study challenging to reproduce elsewhere, the Asian metropolis has several sustainability and circularity programs that can be easily replicable in other contexts.

References

Akande, A., Cabral, P., Gomes, P., & Casteleyn, S. (2019). The Lisbon ranking for smart sustainable cities in Europe. *Sustainable Cities and Society, 44*, 475–487.

Allam, Z., Moreno, C., Chabaud, D., & Pratlong, F. (2021). *Proximity-based planning and the "15-Minute City": A sustainable model for the city of the future* (pp. 1–20). The Palgrave Handbook of Global Sustainability.

Allen, C., Metternicht, G., & Wiedmann, T. (2018). Initial progress in implementing the Sustainable Development Goals (SDGs): A review of evidence from countries. *Sustainability Science, 13*(5), 1453–1467.

Bătăgan, L. (2011). Smart cities and sustainability models. *Informatica Economica, 15*(3), 80–87.

Bibri, S. E. (2018). *Smart sustainable cities of the future*. Springer.

Bibri, S. E., & Krogstie, J. (2017). Smart sustainable cities of the future: An extensive interdisciplinary literature review. *Sustainable Cities and Society, 31*, 183–212.

Bruzzone, M., Dameri, R. P., & Demartini, P. (2021). Resilience reporting for sustainable development in cities. *Sustainability, 13*(14), 7824.

Buffel, T., Phillipson, C., & Scharf, T. (2012). Ageing in urban environments: Developing 'age-friendly' cities. *Critical Social Policy, 32*(4), 597–617.

Capgemini. (2020). *Street smart: Putting the citizen at the center of smart city initiatives*.

Ciacci, A., Ivaldi, E., & González-Relaño, R. (2021). A partially non-compensatory method to measure the smart and sustainable level of Italian municipalities. *Sustainability, 13*, 435.

cordis.europa.eu. (n.d.). *Soluzioni basate sulla natura: Trasformare le città, aumentare il benessere. Récupéré sur CORDIS*. https://cordis.europa.eu/article/id/421853-nature-based-solutions/it

Dameri, R. P. (2013). Searching for smart city definition: A comprehensive proposal. *International Journal of computers & technology, 11*(5), 2544–2551.

Dijkstra, L., Florczyk, A. J., Freire, S., Kemper, T., Melchiorri, M., Pesaresi, M., & Schiavina, M. (2018). *Applying the degree of urbanisation to the globe: A new harmonised definition reveals a different picture of global urbanisation*. IAOS Conference.

EEA. (2021). *Nature-based solutions in europe: Policy, knowledge and practice for climate change adaptation and disaster risk reduction*. European Environment Agency (EEA).

EMF. (2019). *Circular cities: Thriving, liveable, resilient*. Récupéré sur Ellen Mcarthur Foundation. https://ellenmacarthurfoundation.org/topics/cities/overview

EU. (2021). *European commission, directorate-general for research and innovation—evaluating the impact of nature-based solutions: A summary for policy makers*. Publications Office.

Fujita, T. (2009). Eco-towns for 3R promotion in Japan. In *Inaugural meeting of the Regional 3R forum in Asia* (pp. 11–12).

Harper, D. (2020). *Smart cities*. Dans Perspectives from our Asia Pacific Junior Talents (pp. 22–24).

Ishida, T. (2002). Digital city kyoto. *Communications of the ACM, 45*(7), 76–81.

Ivaldi, E., Penco, L., Isola, G., & Musso, E. (2020). Smart sustainable cities and the urban knowledge-based economy: A NUTS3 level analysis. *Social Indicators Research, 150*(1), 45–72.

Johnson, M. P. (2001). Environmental impacts of urban sprawl: A survey of the literature and proposed research agenda. *Environment and Planning A, 33*(4), 717–735.

Kourtit, K., Nijkamp, P., & Arribas, D. (2012). Smart cities in perspective – A comparative European study by means of self-organizing maps. *Innovation: The European Journal of Social Science Research, 25*(2), 229–246.

Laguerre, M. S. (2005). *The digital city* (Vol. 10, p. 9780230511347). Palgrave MacMillan.

Lai, C. S., Jia, Y., Dong, Z., Wang, D., Tao, Y., & Lai, Q. H. (2020). A review of technical standards for smart cities. *Clean Technologies, 2*(3), 290–310.

Lu, P., & Stead, D. (2013). Understanding the notion of resilience in spatial planning: A case study of Rotterdam, The Netherlands. *Cities, 35*, 200–212.

Maldini, I. (2021). *The Amsterdam Doughnut: Moving towards "strong sustainable consumption" policy?* t 4th PLATE 2021 Virtual Conference.

Matos, F., Vairinhos, V. M., Dameri, R. P., & Durst, S. (2017). Increasing smart city competitiveness and sustainability through managing structural capital. *Journal of Intellectual Capital, 18*(3), 693–707.

Maunier, R. (1910). The definition of the city. *American Journal of Sociology, 15*(4), 536–548.

McDonald, R. I. (2008). Global urbanization: Can ecologists identify a sustainable way forward? *Frontiers in Ecology and the Environment, 6*(2), 99–104.

Meini, L., Facchini, A., and Papa, C. (2019, gennaio 22). *Cities of tomorrow: The circular cities*. Récupéré sur Ispi. https://www.ispionline.it/it/pubblicazione/cities-tomorrow-circular-cities-22057

Melkunaite, L., and Guay, F. (2016). *Resilient city: Opportunities for cooperation*. IAIA16 Conference Proceedings, Resilience and Sustainability 36th Annual Conference of the International Association for Impact Assessment, (pp. 11–14).

Moreno, C. (2020). *Droit de cité*. Editions de l'Observatoire.

Moreno, É., & Ochoa, F. A. (2011). Turismo sostenible, cadena de valor y participación comunitaria en Suesca (Cundinamarca), Colombia. *Turismo y Sociedad, 12*, 197–214.

Mueller, N., Rojas-Rueda, D., Khreis, H., Cirach, M., Andrés, D., Ballester, J., et al. (2020). Changing the urban design of cities for health: The superblock model. *Environment International, 134*, 105132.

Newman, P., Beatley, T., & Boyer, H. (2009). Resilient cities: Responsing to peak oil and climate change. *Australian Planner, 46*(1). https://doi.org/10.1080/07293682.2009.9995295

OECD. (2015). *Ageing in cities*. OECD Publishing.

Pancholi, S., Yigitcanlar, T., & Guaralda, M. (2015). Public space design of knowledge and innovation spaces: Learnings from Kelvin Grove Urban Village, Brisbane. *Journal of Open Innovation: Technology, Market, and Complexity, 1*(1), 13.

Papa, R., Galderisi, A., Vigo Majello, M. C., & Saretta, E. (2015). Smart and resilient cities. A systemic approach for developing cross-sectoral strategies in the face of climate change. *TEMA Journal of Land Use, Mobility and Environment, 8*(1), 19–49.

Penco, L., Ivaldi, E., Bruzzi, C., & Musso, E. (2020). Knowledge-based urban environments and entrepreneurship: Inside EU cities. *Cities, 96*, 102443.

Penco, L., Ivaldi, E., Isola, G., & Musso, E. (2019). Business management theories and practices in a dynamic competitive environment. In *Smart sustainable city and urban knowledge-based economy: Evidences from Italy* (pp. 948–966).

PepeResearch. (2022). *Italiani e Smart City–report di ricerca*.

Pickett, S. T. A., Cadenasso, M. L., & Grove, J. M. (2004). Resilient cities: Meaning, models, and metaphor for integrating the ecological, socio-economic, and planning realms. *Landscape and Urban Planning, 69*(4), 369–384.

Pinto, F., & Akhavan, M. (2022). Scenarios for a post-pandemic city: Urban planning strategies and challenges of making "Milan 15-minutes city". *Transportation research procedia, 60*, 370–377.

Rahmat, H. K., Widana, I. D., Basri, A. S., & Musyrifin, Z. (2021). Analysis of potential disaster in the new capital of Indonesia and its mitigation efforts: A qualitative approach. *Disaster Advances, 14*(3), 40–43.

Raworth, K. (2017a). *Doughnut economics: Seven ways to think like a 21st-century economist*. Chelsea Green Publishing.

Raworth, K. (2017b). Why it's time for doughnut economics. *IPPR Progressive Review, 24*(3), 216–222.

Rueda, S. (2019). Superblocks for the design of new cities and renovation of existing ones: Barcelona's case. In D. M. Nieuwenhuijsen & H. Khreis (Eds.), *Integrating human health into urban and transport planning (pp. 135-153)*. Springer International Publishing AG.

Slavova, M., & Okwechime, E. (2016). African smart cities strategies for agenda 2063. *Africa Journal of Management, 2*(2), 210–229.

Tang, Z., Jayakar, K., Feng, X., Zhang, H., & Peng, R. X. (2019). Identifying smart city archetypes from the bottom up: A content analysis of municipal plans. *Telecommunications Policy, 43*(10), 101834.

TDLC. (2017). *Tokyo development learning center–Yokohama: Reinventing the future of a city*. World Bank.

UN. (2018). *World urbanization prospects: The 2018 revision*. UN Department of Economic and Social Affairs Population Division.

UN. (2022). *Make cities and human settlements inclusive, safe, resilient and sustainable*. Récupéré sur https://sdgs.un.org/goals/goal11

UNISDR. (2012). *How to make cities more resilient – A handbook for local government leaders*. Available at: https://www.undrr.org/publication/how-make-cities-more-resilient-handbook-local-government-leaders-2012

van Hoof, J., & Kazak, J. K. (2018). Urban ageing. *Indoor and Built Environment, 27*(5), 583–586.

WHO. (2007). *World Health Organization (2007a) Global age-friendly cities: A guide*. WHO Press.

Winters, J. V. (2011). Why are smart cities growing? Who moves and who stays. *Journal of Regional Science, 51*(2), 253–270.

WorldBank. (2020). *Urban development*. Récupéré sur The World Bank. https://www.worldbank.org/en/topic/urbandevelopment/overview#1

Yigitcanlar, T. (2010). Making space and place for the knowledge economy: Knowledge-based development of Australian cities. *European Planning Studies, 18*(11), 1769–1786.

Chapter 7
Conclusions

Andrea Ciacci and Enrico Ivaldi

Abstract This concluding chapter aims to take stock of the topics explored in this book. As a whole, this book sheds light on the knowledge-based economy and smart and sustainable cities (SSCs) as two interrelated concepts. Specifically, the first chapter analyzes the evolutionary trend of cities towards knowledge, smartness, and sustainability, and identifies successful paradigms of SSCs involving human centricity, service orientation, digitization, innovation, sustainable infrastructural development, natural environment protection, and participative public policies. The second chapter leverages knowledge spillover theories to explore the relationships between a knowledge-based economy and SSCs. The third chapter discusses how entrepreneurial ecosystems based on digital platforms stimulate knowledge flows within them, benefiting a market economy. The fourth chapter defines four indexes to measure the different dimensions of SSCs. The last chapter examines the main features of future cities, from their origins to their evolutionary trajectory, and highlights the need for a profound change in the structure of cities to adequately respond to current challenges. The work also reviews previous studies on the drivers and barriers of urban development and emphasizes the importance of governance and stakeholder involvement in the process of urban transformation.

Keywords Knowledge-based economy · Smart and sustainable city · Technology · Innovation

This work provided a great opportunity to continue the debate on the knowledge-based economy in smart and sustainable cities (SSCs) by embracing this broad topic from different perspectives. The key enabler of the urban paradigm shift is technology. Cities are evolutionary entities that address the current needs of the urban community, such as climate change, overpopulation, malnutrition, overexploitation of natural resources, and air pollution. Innovation and technological advancement facilitate the city's transition and make urban community life more habitable.

The first chapter of this book analyzes how the evolutionary trend of cities toward the pillars of knowledge, smartness, and sustainability. The so-called smart sustainable knowledge city is an open ecosystem created to respond to the most challenging

E. Ivaldi, A. Ciacci, *Smart Sustainable Cities and Knowledge-Based Economy*, Human Well-Being Research and Policy Making,
https://doi.org/10.1007/978-3-031-25038-5_7

issues of urban societies. These cities balance social cohesion, environmental sustainability, and high levels of quality of life to satisfy the needs of the society. SSCs have many factors of attraction and retention for knowledge workers and companies. This attractiveness triggers mechanisms of mutual influence between SSC and knowledge economy. Attracting knowledge workers means developing the creative engine of a city. The chapter identifies paradigms of successful smart and sustainable knowledge cities. They show many similarities involving the pillars of human centricity, service orientation, digitization, innovation, sustainable infrastructural development, natural environment protection, and participative public policies.

The second perspective of analysis defines the relationships between a knowledge-based economy and SSCs by leveraging knowledge spillover theories. The knowledge economy and SSCs influence each other. Many actors and organizations in a knowledge-based economy participate in this process of value creation for SSCs. These interconnected relationships between a knowledge-based economy and SSCs create patterns of mutual influence as a knowledge transmission circuit. In this regard, entrepreneurship plays a crucial role as it enables the commercialization of knowledge. Entrepreneurship plays an active role in determining the spread of knowledge and the formation of new businesses. A city must leverage its pull factors to increase its attractiveness in the eyes of knowledge workers. "People-first" strategies, management of "glocal tensions" (e.g., balancing global knowledge creation with local production), and a vision focused on sustainability through improved smartness represent the strategic pillars identified in the literature.

The third perspective of analysis argues and discusses how entrepreneurial ecosystems based on digital platforms stimulate knowledge flows within them benefiting in favor of a market economy. Digital platforms enhance the sustainability of cities by facilitating knowledge exchanges with partners and improving knowledge management processes within the enterprise. Public-private partnerships should incentivize the adoption of digital platforms for collaborative purposes. In turn, businesses should strive to develop IT capabilities and incorporate digital platforms into their business models to ensure they function at scale. The great potential of business models powered by digital platforms and inter-organizational collaborations can benefit society as a whole. Finally, smarter and more innovative cities also attract human capital when this multilevel synergy generates positive returns for all stakeholders. Innovation must meet social needs through the control of public decision-makers, while private actors generate value for society through their innovations while ensuring adequate revenues for city revitalization.

The fourth perspective of analysis examines the definition of four indexes to measure the different dimensions of SSCs. Statistical applications show a high correlation between the rankings and how the choice of a specific method does not produce significantly different results from those we would have obtained by applying an alternative method, thus indicating a convergence toward the same results. It is thus confirmed that the methodological choice should therefore depend on alignment with the general theoretical framework and not depend on different computational methodologies.

The last analysis concerns the definition of the main features of future cities, from their origins to their evolutionary trajectory highlighting the need for a profound change in the structure of cities to adequately respond to current challenges. From the original concept of a "digital city" to the final concept of a "circular city," moving on to "smart," "sustainable," "resilient," and "15-minute" cities. There are three pillars of the new circular city that administrations must invest in: buildings, mobility, and products highlighting the importance that urban changes of this kind entail for the world we live in. An overall future transition to this type of city cannot be separated from the final analysis of the available drivers and barriers that local governments face.

Schuch de Azambuja (2021) conducted a systematic literature review about the factors enabling and hindering the development of cities: the analysis took into consideration 169 scientific articles, highlighting a high number of drivers and barriers that have been clustered in 5 reference macro-areas (society, environment, economy, governance, and urban infrastructure). The results of this survey indicated the governance factor as the priority in which to concentrate resources for the development of a city, on the contrary, the most critical element turned out to be the involvement of stakeholders in the process of urban transformation.

Ferraris et al. (2020) focused on the theme of "open innovation" and how public administrations should take advantage of this tool to improve the lives of citizens. Chesbrough et al. (2006) defined the concept of open innovation as follows: "a paradigm that assumes that firms can and should use external ideas as well as internal ideas, and internal and external paths to market, as the firms look to advance their technology." On the basis of this assumption, public administrations should acquire knowledge and technologies from outside (private companies) in order to accelerate the innovation process. This study conducted a series of interviews directly with stakeholders to understand how city management should operate in order to create a stimulating business environment conducive to public-private partnerships.

Veselitskaya et al. (2019) conducted research on enabling and limiting factors for the development of cities through the use of four case studies: the cities of Barcelona (Spain), Charlotte (United States), Shanghai (China), and Tokyo (Japan). The results of this study show that the main critical issues are related to the conflict of interest between municipalities, population, businesses, and cyber security issues.

In Table 7.1 some of the enabling and hindering factors of city development are reported.

The heralded rise of "smart knowledge cities" was expected to bring data-centric solutions to urban challenges. While Asia remains at the forefront, traditional cities are now feeling the pressure to upgrade aging infrastructure. The COVID-19 pandemic, growing commitments to sustainability and circularity, resource constraints, and continued urban growth make new investments necessary. It has never been

Table 7.1 Drivers and barriers to city development. Source: authors' elaboration based on (Ferraris et al., 2020; Veselitskaya et al., 2019; Schuch de Azambuja, 2021)

Authors	Drivers	Barriers
Schuch de Azambuja, (2021)	• Public provision of urban services and innovation • Healthcare and sanitation facilities • Urban attractiveness • Tourist attractive projects • Air pollution monitoring, emission control systems • Smart waste management • Transparency and openness • Physical infrastructure integration	• Lack of citizen participation • Lack of trust • Lack of social awareness • Cultural diversification • Citizen's inequality • Digital divide • Resistance to change • Social exclusion and gentrification • Unavailability of services for different communities • Lack of connection between technological and social infrastructure
Veselitskaya et al., (2019)	• High tourist influx potential • Modern infrastructure • Broad implementation of ICT and mobile solutions • Development of public-private partnerships	• Possibility of inconsistency between the interests of citizens, private capital, and political elites. • Austerity policy and urban dynamics lead to land leases.
Ferraris et al., (2020)		• Lack of rules, tasks, and responsibility • Insufficiently integrated view of the city planning. • Lack of fit of administrative styles and interdepartmental coordination and communication. • Risk aversion • Data availability • Disincentives and non-flexible public procurement rules • Lack of resources • Lack of technological capabilities

more important to make cities smarter, more efficient, and more sustainable for residents.

Never more so than now.

Acknowledgement The authors would like to thank Marianna Bartiromo and Tiziano Pavanini for their invaluable assistance in the drafting of this volume. We also thank the editors, M. Joseph Sirgy and Richard J. Estes, for giving us the opportunity to contribute to the series "Human Well-Being Research and Policy Making" which allowed us to reflect and increase our knowledge on a cutting-edge topic.

References

Chesbrough, H., Vanhaverbeke, W., & West, J. (2006). *Open innovation: Researching a new paradigm*. Oxford University Press on Demand.

Ferraris, A., Santoro, G., & Pellicelli, A. C. (2020). "Openness" of public governments in smart cities: Removing the barriers for innovation and entrepreneurship. *International Entrepreneurship and Management Journal, 16*(4), 1259–1280.

Schuch de Azambuja L. (2021). Drivers and barriers for the development of smart sustainable cities: A systematic literature review. *14th International Conference on Theory and Practice of Electronic Governance*, (pp. 422–428).

Veselitskaya, N., Karasev, O., & Beloshitskiy, A. (2019). Drivers and barriers for smart cities development. *Theoretical and Empirical Researches in Urban Management, 14*(1), 85–110.

